Advances in Food and By-Products Processing Towards a Sustainable Bioeconomy

Advances in Food and By-Products Processing Towards a Sustainable Bioeconomy

Special Issue Editors

Nikolaos Kopsahelis
Vasiliki Kachrimanidou

MDPI • Basel • Beijing • Wuhan • Barcelona • Belgrade

MDPI

Special Issue Editors
Nikolaos Kopsahelis
Ionian University
Greece

Vasiliki Kachrimanidou
The University of Reading
UK

Editorial Office
MDPI
St. Alban-Anlage 66
4052 Basel, Switzerland

This is a reprint of articles from the Special Issue published online in the open access journal *Foods* (ISSN 2304-8158) in 2019 (available at: https://www.mdpi.com/journal/foods/special_issues/Advances_Food_By-Products_Processing_Towards_Sustainable_Bioeconomy)

For citation purposes, cite each article independently as indicated on the article page online and as indicated below:

LastName, A.A.; LastName, B.B.; LastName, C.C. Article Title. *Journal Name* **Year**, *Article Number*, Page Range.

ISBN 978-3-03921-752-6 (Pbk)
ISBN 978-3-03921-753-3 (PDF)

Contents

About the Special Issue Editors

Nikolaos Kopsahelis is a Chemist (University of Patras, Greece, 2003) with an MSc in Food Biotechnology (University of Ulster, UK, 2004) and PhD in Chemistry (Food Chemistry and Biotechnology, University of Patras, Greece, 2009). He is currently Associate Professor (Food Chemistry & Industrial Fermentations) at the Department of Food Science & Technology (Kefalonia Campus, Ionian University). His research focuses on fermentation technology in food (including agricultural products and wastes) chemistry and analysis, the biotechnological valorization of renewable resources, and in the development of novel biorefineries. He has participated as a member of research teams in 14 research projects funded by the EU, as well as National and private funds, and is currently involved in 7 research projects, where he is the coordinator in 3 of these projects. He has published 50 papers (mean IF > 5.6) in international peer-reviewed journals (>2200 citations), 4 book chapters, 2 patents, and has 68 publications in international and national conference proceedings.

Vasiliki Kachrimanidou is an Agronomist/Food Scientist, graduated from the Department of Food Science and Technology in the Agricultural University of Athens. Her PhD thesis (Department of Food Science & Human Nutrition, Agricultural University of Athens) focused on the development of biorefinery concepts for the microbial production of poly(hydroxyalkanoates) and value-added products using byproduct streams from sunflower-based biodiesel production. She has worked as a postdoctoral research associate at the Agricultural University of Athens as well as the University of Reading, and is currently based at the Department of Food Science and Technology (Ionian University). Her research interests are focused on the development of integrated biorefinery concepts using agroindustrial waste streams and byproducts to generate value-added products towards the transition to circular economy. Up to date, she has 18 publications in international peer reviewed journals and 4 book chapters, along with numerous contributions in international conferences.

![foods logo] **foods**

MDPI

Editorial

Advances in Food and Byproducts Processing towards a Sustainable Bioeconomy

Nikolaos Kopsahelis [1,*] and Vasiliki Kachrimanidou [1,2,*]

[1]   Department of Food Science and Technology, Ionian University, 28100 Argostoli, Kefalonia, Greece
[2]   Department of Food and Nutritional Sciences, University of Reading, Reading RG6 6AP, UK
*   Correspondence: kopsahelis@upatras.gr (N.K.); vkachrimanidou@gmai.com (V.K.)

Received: 6 September 2019; Accepted: 18 September 2019; Published: 19 September 2019

Keywords: biorefineries; circular economy; bioeconomy; food processing; food biotechnology; bioprocesses; bioactive compounds; food waste valorization; sustainability

---

The bioeconomy concept was initially focused on resource substitution, aiming to mitigate the depletion of fossil resources and confer an alternative approach for resource utilization. Within this context, the production of biomass from diversified resources, along with its subsequent conversion, fractionation, and processing, was deemed of paramount importance. Thus, biotechnological and chemical routes along with process engineering were implemented for the production and marketing of food, feed, fuel, and fiber.

Nowadays, although resource substitution is still considered important, and regardless of the substantial accomplishments of chemical treatments for efficient biomass utilization, emphasis has shifted towards the biotechnological innovation perspective—to configure bioprocesses that will be included in the restructure of current facilities to develop consolidated processes, under the frame of transition to a circular model that will confer environmental sustainability.

Alongside this shift, global projections of food loss are estimated at one-third of the total quantity produced for human consumption, posing not only a sustainability issue related to food security but also a significant environmental concern. Likewise, food waste streams, derived primarily from fruits and vegetables, cereals, oilseeds, meat, dairy, and fish processing, constitute an abundant pool of complex carbohydrates, proteins, lipids, and functional compounds. Hence, the deployment of food waste streams as raw materials will encompass the formulation of added-value products that will be ideally reintroduced in the food supply chain to close the loop.

This Special Issue is devoted to the development of innovative and emerging food and byproducts processing methods, as a necessity for the sustainable transition to a bioeconomy era. Valorization, bioprocessing and biorefining of food-industry-based streams, the isolation of high added-value compounds, applications of the resulting bio-based chemicals in food manufacturing, novel food formulations, economic policies for food waste management, along with sustainability or technoeconomic assessment of processing methods constitute subject areas that need to be addressed.

More specifically, bioprocess design to valorize food-industry waste and byproducts streams should be initiated by characterizing the composition of the onset raw material with the aim of identifying the target end-products, whereas the generation of multiple high added-value products is a prerequisite for cost-effective processes to establish economic sustainability. On top of that, the feasibility of innovative processing could be sustained by encompassing food applications, driven by the constantly emerging consumers' demand for functional foods with enhanced nutritional value. Equally, a growing awareness for bio-based and natural food components is being developed, thereby imposing challenges on the substitution of chemically derived ingredients with their natural counterparts.

Within this context, Papadaki et al. [1] evaluated the production of *Morchella* sp. as one of the most expensive mushrooms conferring high nutritional value, using agroindustrial starch-based substrates, specifically wheat grains, potato peels, and a mixture of both. Submerged and solid-state fermentations were employed to study biomass formation and polysaccharide content. In another study, Koliastasi et al. [2] presented the utilization of composted olive processing solid waste to extract emulsifiers, composed mainly of oligosaccharides and oligopeptides. The emulsifying capacity and stabilizing properties of the extracted emulsifiers indicated that valorization of olive processing waste through partial-composting might provide a novel, fast, and economic method of production of high added-value products. Likewise, Hou et al. [3] presented the effect of drying methods (dehydration freeze-drying) on the properties of heat-induced gels using porcine plasma protein powder.

Apart from applications in food formulation, high-value products as food packaging materials can also be obtained using renewable sources. A case examining the interaction of arabinoxylan films with soil and water was provided by Anderson and Simsek [4]. Wheat bran, maize bran, and dried distillers grain combined with glycerol or sorbitol were used to formulate arabinoxylan films, followed by the study of biodegradability, water solubility, and water vapor transmission rate.

The employment of bacterial, fungal, and yeast strains to yield added-value products via their proliferation on renewable substrates comprises the fundamental principle of bioprocess design. Likewise, enzymes entail a frequently studied end-product in fermentative bioconversions. Makanjuola et al. [5] used sorghum bran, a processing waste rich in starch, for the secretion of glucoamylase in submerged fermentations using the fungal strain *Aspergillus awamori*. Cultivation parameters were evaluated to obtain maximum enzyme production, and subsequently hydrolysis experiments were performed, resulting in a glucose-rich solution for further involvement in microbial fermentation.

Among the most studied bio-based products, microbial oil synthesized by oleaginous yeasts and fungi has attracted extensive research interest for the subsequent implementation for biodiesel production or the inclusion in food formulation. Patel et al. [6] studied the culture of *Rhodosporidium kratochvilovae* on clarified butter sediment waste for the production of single cell oil, demonstrating high lipid accumulation. The compliance with ASTM 6751 and EN 14214 international standards was also evaluated, indicating the potential use for biodiesel synthesis. In a similar study, Tsakona et al. [7] proposed the utilization of diversified mixed confectionery waste streams in a two-stage bioprocess to generate nutrient rich feedstock for the proliferation of *Rhodosporidium toruloides* DSM 4444 to generate microbial oil under various fermentation strategies. The fatty acid profile was altered based on the initial substrate, with oleic acid being the predominant one, thereby insinuating the production of tailor-made lipids targeting innovative food applications.

Ultimately, the transition to a bioeconomy era will be facilitated by designing consolidate bioprocesses that confer a holistic approach and focus on innovative end-applications, which could be integrated in existing manufacturing plants. A holistic and efficient resource utilization is unequivocally a prerequisite to consider within the development of fermentation bioconversions. For instance, the proposed biorefinery scheme using diversified mixed confectionery streams suggested the generation of value-added products that could be reintroduced in the food supply chain. In a similar manner, Lappa et al. [8] elaborated the most recent advances and approaches to benefit from the complete exploitation of cheese whey stream. The proposed integrated biorefineries target the cost-effective production of novel value-added products to convey enhanced sustainability and incorporate a "zero waste" approach. Retrospectively, the design of consolidate bioprocesses towards a circular and bio-based economy should also take into account the founding pillars of economy, society, and environment. It is envisaged that this Special Issue will succeed in providing state-of-the-art advances in food and byproducts processing within the bioeconomy era.

Acknowledgments: We acknowledge support of this work by the project "Research Infrastructure on Food Bioprocessing Development and Innovation Exploitation – Food Innovation RI" (MIS 5027222), which is implemented under the Action "Reinforcement of the Research and Innovation Infrastructure", funded by the Operational Programme "Competitiveness, Entrepreneurship and Innovation" (NSRF 2014-2020) and co-financed by Greece and the European Union (European Regional Development Fund).

Conflicts of Interest: The authors declare no conflict of interest.

References

1. Papadaki, A.; Diamantopoulou, P.; Papanikolaou, S.; Philippoussis, A. Evaluation of Biomass and Chitin Production of Morchella Mushrooms Grown on Starch-Based Substrates. *Foods* **2019**, *8*, 239. [CrossRef] [PubMed]
2. Koliastasi, A.; Kompothekra, V.; Giotis, C.; Moustakas, A.K.; Skotti, E.P.; Gerakis, A.; Kalogianni, E.; Ritzoulis, C. Emulsifiers from Partially Composted Olive Waste. *Foods* **2019**, *8*, 271. [CrossRef] [PubMed]
3. Hou, C.; Wang, W.; Song, X.; Wu, L.; Zhang, D. Effects of Drying Methods and Ash Contents on Heat-Induced Gelation of Porcine Plasma Protein Powder. *Foods* **2019**, *8*, 140. [CrossRef] [PubMed]
4. Anderson, C.; Simsek, S. How Do Arabinoxylan Films Interact with Water and Soil? *Foods* **2019**, *8*, 213. [CrossRef] [PubMed]
5. Makanjuola, O.; Greetham, D.; Zou, X.; Du, C. The Development of a Sorghum Bran-Based Biorefining Process to Convert Sorghum Bran into Value Added Products. *Foods* **2019**, *8*, 279. [CrossRef] [PubMed]
6. Patel, A.; Sartaj, K.; Pruthi, P.A.; Pruthi, V.; Matsakas, L. Utilization of Clarified Butter Sediment Waste as a Feedstock for Cost-Effective Production of Biodiesel. *Foods* **2019**, *8*, 234. [CrossRef]
7. Tsakona, S.; Papadaki, A.; Kopsahelis, N.; Kachrimanidou, V.; Papanikolaou, S.; Koutinas, A. Development of a Circular Oriented Bioprocess for Microbial Oil Production Using Diversified Mixed Confectionery Side-Streams. *Foods* **2019**, *8*, 300. [CrossRef] [PubMed]
8. Lappa, I.K.; Papadaki, A.; Kachrimanidou, V.; Terpou, A.; Koulougliotis, D.; Eriotou, E.; Kopsahelis, N. Cheese Whey Processing: Integrated Biorefinery Concepts and Emerging Food Applications. *Foods* **2019**, *8*, 347. [CrossRef] [PubMed]

foods

MDPI

Article

Development of a Circular Oriented Bioprocess for Microbial Oil Production Using Diversified Mixed Confectionery Side-Streams

Sofia Tsakona [1], Aikaterini Papadaki [1,2,*], Nikolaos Kopsahelis [2], Vasiliki Kachrimanidou [3], Seraphim Papanikolaou [1] and Apostolis Koutinas [1,*]

[1] Department of Food Science and Human Nutrition, Agricultural University of Athens, Iera Odos 75, 11855 Athens, Greece
[2] Department of Food Science and Technology, Ionian University, 28100 Argostoli, Greece
[3] Department of Food and Nutritional Sciences, University of Reading, Reading RG6 6AD, UK
*   Correspondence: kpapadaki@aua.gr (A.P.); akoutinas@aua.gr (A.K.)

Received: 23 June 2019; Accepted: 29 July 2019; Published: 31 July 2019

Abstract: Diversified mixed confectionery waste streams were utilized in a two-stage bioprocess to formulate a nutrient-rich fermentation media for microbial oil production. Solid-state fermentation was conducted for the production of crude enzyme consortia to be subsequently applied in hydrolytic reactions to break down starch, disaccharides, and proteins into monosaccharides, amino acids, and peptides. Crude hydrolysates were evaluated in bioconversion processes using the red yeast *Rhodosporidium toruloides* DSM 4444 both in batch and fed-batch mode. Under nitrogen-limiting conditions, during fed-batch cultures, the concentration of microbial lipids reached 16.6–17 g·L^{-1} with the intracellular content being more than 40% (w/w) in both hydrolysates applied. *R. toruloides* was able to metabolize mixed carbon sources without catabolite repression. The fatty acid profile of the produced lipids was altered based on the substrate employed in the bioconversion process. Microbial lipids were rich in polyunsaturated fatty acids, with oleic acid being the major fatty acid (61.7%, w/w). This study showed that mixed food side-streams could be valorized for the production of microbial oil with high unsaturation degree, pointing towards the potential to produce tailor-made lipids for specific food applications. Likewise, the proposed process conforms unequivocally to the principles of the circular economy, as the entire quantity of confectionery by-products are implemented to generate added-value compounds that will find applications in the same original industry, thus closing the loop.

Keywords: food-processing; circular economy; bioprocess development; *Rhodosporidium toruloides*; microbial oil; oleic acid

1. Introduction

The concept of the circular economy is emerging as a worldwide strategy to transit from the current linear economy model of production and consumption to efficient resource exploitation [1,2]. Within this framework, bio-economy encompasses the holistic valorization of renewable resources towards the development of biorefinery concepts and bioprocessing schemes to produce high value-added products. Evidently, microbial lipid production constitutes a research area of paramount significance. In particular, the production of highly unsaturated microbial lipids has gained the attention of many researchers, as they could be widely employed in functional food formulations, eliciting high nutritional composition [3].

Microbial lipids are secondary metabolites synthesized by oleaginous yeasts, fungi, and algae, exhibiting an intracellular accumulation of more than 20% [4]. Nonetheless, the industrial production of microbial oil impairs numerous impediments associated with the downstream separation of intracellular lipids, carbon source, and operating cost of the fermentative process. Actually, an economic assessment of large-scale fermentation of *R. toruloides* using glucose resulted in an estimated unitary production cost of $3.4·kg^{-1} at an annual production capacity of 10,000 t [5]. Hence, to establish an economically feasible and sustainable process, it is imperative to fulfill certain criteria, including high productivity, the development of biorefinery concepts that generate multiple high-value end-products, or a high-end market price of the formulated product. For instance, Kopsahelis et al. [6] presented a biorefinery process suggesting the simultaneous production of protein isolate, tartrates, ethanol, polyphenols, and microbial oil from wine lees and cheese whey. Ochsenreither et al. [3] speculated that commercialization could only be achieved via the manufacture of high-value products, such as the addition of polyunsaturated fatty acids in food applications. It is also unequivocal that the cost of the initial on-set material should be negligible; thus, waste and by-products streams exhibit ideal zero or low-cost substrates for the fermentative production of microbial oil.

Among the various oleaginous microorganisms, *Rhodosporidium* species have been widely investigated for microbial oil production via the valorization of an ample range of carbon sources, including glucose, glycerol, fructose, inulin, and xylose, among them [7]. Specifically, *R. toruloides* exhibits advantages over other oleaginous yeasts, including rapid proliferation and high lipid accumulation on low-cost resources [8]. Leiva-Candia et al. [9] reported a microbial oil production ranging 18.1–19.2 g·L^{-1} using different sunflower meal hydrolysates, deriving from the fractionation of sunflower meal, along with crude glycerol as fermentation supplements. Likewise, *R. toruloides* was recently employed for microbial oil production, using a nutrient supplement deriving from flour-rich waste streams (FRW) and wheat milling by-products [10]. Actually, authors proposed a process to contribute towards the valorization of the massive amounts of annual waste streams that occur from confectionery manufacturing industries and bakeries, or equally as discarded, damaged, or out of date products that return on site. Current waste treatments for these streams include animal feed, composting, or disposal in landfills. The process proposed as an alternative option by Tsakona et al. employed a two-stage bioprocess, where high final lipid concentrations and conversion yields were achieved [10,11]. However, only part of the confectionery waste streams (flour-rich streams) was evaluated in the fermentation process. Besides, it is a fact that the majority of studies, dealing with microbial lipids, have been mostly applied for biodiesel and biolubricants production [3,9,12]. Nevertheless, in the context of the circular economy, some recent studies focused on the valorization of food side-streams for the development of microbial oil-based food additives, such as wax esters [13] or directed specifically on the synthesis of high oleic acid microbial lipids [14,15]. Oleic acid resulted in more than the US $350 million in revenues for 2016, whereas a constant increase is projected based on the chemical industry applications [16]. On top of that, the possibility to modulate the fractions of oleic acid in the produced lipids would enable the production of tailor-made microbial lipids for specifically targeted applications [8]. It is thus imperative that in the frame of the circular economy, food waste streams should emerge as a potential feedstock for the synthesis of microbial lipids, targeting special food applications. For instance, oleogels deriving from high unsaturated vegetable oils are desirable for the substitution of trans fat content in foods [17], which has been recently banned by U.S. Food and Drug Administration (FDA) [18].

The aim of this study was the consolidated valorization of diversified confectionary waste streams, rich in mixed carbohydrates and other micronutrients for the fermentative production of microbial lipids using the oleaginous red yeast *R. toruloides*. The present study constitutes a more integrated extension of our previous work [10,11], that implemented only part of the confectionery waste streams (flour-rich waste streams) in the investigation during the hydrolysis and fermentation process. Nonetheless, it is crucial to configure a bioprocess that will exploit the full potential of the confectionery waste streams, to generate value-added products that can be reintroduced in the food supply chain under the context of zero waste and enhanced sustainability. More specifically, the present study targeted the valorization of all different confectionery waste streams (containing sucrose, starch, and lactose, among others), which was not previously reported, towards the development of a holistic cascade bioprocess. The effect of the different confectionery side-streams was evaluated on the ability of *R. toruloides* to metabolize confectionary hydrolysates and shift the carbon flux towards lipid synthesis under nitrogen limitation conditions. The fatty acid profile was also evaluated, aiming to identify tailor-made food applications. Interestingly, valorization in a two-stage bioprocess of mixed confectionery waste streams, as described in the present study, entailed modifications in the composition of microbial oil. A higher degree of unsaturated lipids was obtained, advocating the potential to enhance the feasibility of the proposed scheme for further integration in existing facilities, targeting the development of high value-added products, which under the frame of the circular economy might be applied in food formulation within the initial industry.

2. Materials and Methods

2.1. Microorganisms

The fungal strain *Aspergillus awamori* 2B.361 U2/1 that was originally obtained from ABM Chemicals, Ltd. (Woodley, UK), and was kindly provided by Professor Colin Webb (University of Manchester, Manchester, UK), was employed for the production of crude enzyme consortia during solid-state fermentation (SSF). Storage, maintenance, sporulation, and inoculum preparation of the fungal strain *A. awamori* 2B.361 U2/1 have been previously reported by Tsakona et al. [11].

The oleaginous yeast strain *Rhodosporidium toruloides* DSM 4444 was used in the fermentative production of microbial oil. The strain was maintained at 4 °C, on slopes containing glucose (10 g·L^{-1}), yeast extract (10 g·L^{-1}), peptone (10 g·L^{-1}), and agar (2%, w/v). A liquid pre-culture with the same composition was prepared as fermentation inocula.

2.2. Raw Materials

Wheat-milling by-products containing (w/w) 12% starch, 20% protein, and 1.1% phosphorus were employed as the solid substrate in SSF with the fungal strain *A. awamori*. Mixed food for infants (MFI), mixed confectionery waste streams (MCWS), and mixed waste streams (MWS), obtained from different categories of confectionery waste streams, were all supplied by Jotis S.A. (Athens, Greece), a Greek confectionery industry. They were involved in enzymatic hydrolytic reactions, for the formulation of fermentative substrates and subsequent microbial oil production. MFI contained (w/w) 33% starch, 17% sucrose, and 27% lactose. MCWS demonstrated a similar composition (32.3% starch, 16% sucrose, 27% lactose) along with 7% (w/w) of lipids. Lipids were extracted with n-hexane (Sigma-Aldrich, St. Louis, MO, USA) for seven days before utilization. Likewise, the composition of FRW (84.8% starch, 7.3% protein) has been previously described [10]. MWS contained a ratio of 1:1:1 MFI:MCWS:FRW and presented a final composition of (w/w) 50% starch, 11% sucrose, and 18% lactose.

2.3. Solid-State Fermentation and Enzymatic Hydrolytic Experiments

SSFs were conducted in 250 mL Erlenmeyer flasks for the production of crude enzyme consortia, as described by Tsakona et al. [11]. Briefly, 5 g of wheat milling by-products were added in each flask and autoclaved at 121 °C for 20 min. Subsequently, the solids were inoculated with a fungal spore suspension (2 × 10^6 spores·mL^{-1}) of *A. awamori* that was also used to adjust the moisture content of the substrate.

The fermented solids of five flasks (after 3 days of incubation at 30 °C) were suspended in 500 mL sterilized tap water and subsequently macerated using a kitchen blender. After centrifugation (9000× *g* for 10 min), individual hydrolysis of each waste stream (MFI, MCWS, or MWS) was performed by adding the supernatant in 1 L Duran bottles containing known quantities (50, 100, 150 g·L^{-1}) of MFI, MCWS, or MWS. The suspension was mixed employing magnetic stirrers, and enzymatic hydrolysis was carried out at 55 °C and uncontrolled pH conditions.

At the end of enzymatic hydrolysis, the produced hydrolysates were centrifuged (9000× *g* for 10 min), and the supernatant was filter-sterilized using a 0.2 μm filter unit (Polycap TM AS, Whatman Ltd., Maidstone, UK). The pH of the hydrolysate was adjusted to 6, which is optimum for yeast growth, using 5 M NaOH.

2.4. Microbial Oil Fermentations

Shake flask experiments were carried out in 250 mL Erlenmeyer flasks with a working volume of 50 mL, using commercial carbon sources. Carbon sources were selected based on the composition in the hydrolysates obtained after enzymatic reactions, as indicators of the yeast performance in these hydrolysates. More specifically, commercial glucose, sucrose, fructose, and galactose were individually evaluated for yeast proliferation and microbial oil accumulation by *R. toruloides*. Subsequently, shake flask cultivations were performed to study the potential of *R. toruloides* in the hydrolysates of the three waste streams, as described in Section 2.2. All flasks were inoculated with 10% (v/v) of a 24 h exponential pre-culture of *R. toruloides* and incubated at 27 °C in an orbital shaker (ZHWY – 211C Series Floor Model Incubator, Zhicheng, Shanghai, China) using an agitation rate of 180 rpm. The pH value was adjusted during fermentation using 5 M NaOH when needed. Fermentations were carried out in duplicates and the respective analyses in triplicates. Data presented are the mean values of those measurements.

Bioreactor fermentations were conducted in a 3 L bioreactor (New Brunswick Scientific Co., Edison, New Jersey, USA) with a working volume of 1.5 L. The temperature and aeration were set at 27 °C and 1.5 vvm, respectively, whereas pH value was automatically adjusted to 6 with 10 M NaOH. A 10% (v/v) inoculum was applied using a 24 h exponential pre-culture. The agitation rate was controlled in the range of 150–500 rpm to maintain the dissolved oxygen concentration at 20% of saturation.

Fermentations were conducted under fed-batch mode to evaluate the utilization of MFI and MCWS hydrolysates using an optimum carbon to Free Amino Nitrogen (FAN) ratio (C/FAN) based on the results presented by Tsakona et al. [10]. Carbon was calculated based on the carbon content of sugars, whereas the FAN corresponded to the nitrogen contained in the free amino groups of amino acids and peptides in the hydrolysate. The initial FAN concentration was 292–340 mg·L^{-1}, whereas initial total sugar concentration was ~50 g·L^{-1}. Fed-batch fermentation strategy was achieved by the periodic addition of a concentrated solution derived from each hydrolysate (70%, w/v). The feeding solution was added in the bioreactor under aseptic conditions to sustain microbial proliferation and microbial oil synthesis. Production of total dry weight (TDW) and lipid synthesis indicated the termination of the bioprocess, which lasted up to 120 h. Samples were withdrawn to assess sugar and nitrogen consumption along with biomass, lipid, and intracellular polysaccharides (IPS) concentration. Residual cell weight (RCW) was determined by subtracting the produced microbial oil (g·L^{-1}) from TDW (g·L^{-1}).

2.5. Analytical Methods

The analytical methods used in this study were previously described in detail by Tsakona et al. [11]. Briefly, FAN concentration was measured with the ninhydrin colorimetric method, whereas inorganic phosphorus (IP) concentration was analyzed by the ammonium molybdate spectrophotometric method. The concentration of sugars was quantified using a High-Performance Liquid Chromatography unit (Waters 600E, Waters, Milford, MA, USA) equipped with an Aminex HPX-87H column (300 mm × 7.8 mm, Bio-Rad, Hercules, CA, USA) and a differential refractometer (RI Waters 410). Operating conditions were as follows: sample volume 20 µL; mobile phase 0.005 M H$_2$SO$_4$; flow rate 0.6 mL·min^{-1}; column temperature 65 °C. TDW was measured by drying the yeast biomass at 105 °C for 24 h, and microbial oil (MO) was determined according to the method proposed by Folch et al. [19]. Following the disruption of dried yeast cell mass, the Folch solution, chloroform/methanol mixture at a ratio of 2:1 (v/v), was added to the cell debris. The suspension was centrifuged (9000 × *g*, 4 °C, 5 min), the solvent phase was collected, washed with 0.88% KCl (w/v), dried with anhydrous Na$_2$SO$_4$, and evaporated under vacuum. The fatty acid profile of microbial oil was analyzed through the production of fatty acid methyl esters (FAME) following a two-step reaction with methanol using sodium methoxide (MeONa) and HCl as catalysts. FAME was analyzed by a Gas Chromatography Fisons 8060 (Fisons Instruments, Mainz, Germany) unit equipped with a chrompack column (60 m × 0.32 mm) and an FIDdetector. Helium was used as carrier gas (2 mL·min^{-1}). The analysis was carried out at 200 °C with the injector at 240 °C and the detector at 250 °C. The split ratio was 1:50, and the sample injection volume was 1 µL. Peak identification was accomplished by comparison of retention times with those of a certified reference FAME mixture (Supelco 37 Component FAME mix, Sigma-Aldrich, St. Louis, MO, USA). Fatty acid data were expressed as the area percentage of FAME.

The concentration of IPS was calculated using a modified method, described by Liang et al. [20]. Briefly, 50 mg of TDW was treated with 20 mL HCl (2.5 M). Acidification of the suspension was performed at 100 °C for 30 min followed by neutralization to pH 7 with KOH. Samples were filtered (Whatman filter paper), analyzed via the 3,5-dinitrosalicylic acid assay, and IPS were expressed as glucose equivalents, as previously described [11].

3. Results and Discussion

3.1. Enzymatic Hydrolysis of Different Mixed Confectionery Waste Streams

In a previous study, Tsakona et al. presented the production of a nutrient-rich fermentation feedstock deriving from the hydrolysis of FRW using crude enzymatic extracts obtained via SSF with *A. awamori* on wheat milling by-products (WMB) [11]. In this study, a similar approach was employed to hydrolyze mixed and diversified confectionery waste streams, e.g., MFI, MCWS, MWS, therefore, exploiting all potential waste streams.

The aforementioned streams contain not only starch, compared to FRW, but also other sources of carbohydrates, e.g., sucrose and lactose. Thus, the study initially targeted the hydrolysis of all carbohydrate sources into their respective monomeric sugars, to be easily assimilated by microbial entities. Likewise, the hydrolysis of proteins into adequate quantities of amino acids and peptides that can be readily metabolized during a fermentation process was also of high importance. Table 1 presents the degree of hydrolysis of starch, sucrose, and lactose content of all applied waste streams (MFI, MCWS, and MWS). Conversion of starch to glucose reached more than 95% when an initial concentration of 50 g·L^{-1} of MFI was applied. The conversion yield of sucrose to glucose and fructose reached 79.4–91.9%, whereas a similar pattern of lactose conversion to glucose and galactose was observed (75.1–89.2%) for all initial MFI solid concentrations. Starch and sucrose hydrolysis yields of MCWS and MWS (91–96.9% and 78.4–92.1%, respectively) were similar to those of MFI, while lactose conversion yields ranged from 70.1–72.3% for MCWS, and from 73.8–88.9% for MWS. The final sugar composition of the MFI and MCWS hydrolysates presented similar composition. MFI was comprised of 73.4 ± 2.1% glucose, 10.5 ± 1% fructose, and 16.1 ± 2.4% galactose, whereas MCWS contained 74.0 ± 2.7% glucose, 11.1 ± 1.3% fructose, and 14.9 ± 1.6% galactose. MWS hydrolysate contained mainly glucose (83 ± 2.7%) and lower amounts of fructose (6.6 ± 0.7%) and galactose (10.4 ± 0.9%).

The fungal strain *A. awamori* is of high industrial importance for bioprocesses; hence, it has been widely evaluated, particularly, in solid-state fermentation for the production of hydrolytic enzymes [11,21–23]. Smaali et al. reported on the production of extracellular thermostable invertase during submerged cultures, triggered by the addition of sucrose (1%) [23]. Bertolin et al. studied the production of glucoamylase via SSF of *A. awamori* on wheat bran [24]. Grape pomace was employed as the sole substrate for the production of cellulase, xylanase, and pectinase with *A. awamori* 2B.361 U2/1 [25]. The authors reported cellulase activity up to 9.6 ± 0.76 IU/gds during the first 24 h of fermentation. Cellulases comprise three distinct categories, e.g., endo-1,4-β-glucanase, exo-1,4-β-D-glucanase, and β-glucosidase, cleaving the β-1.4 linkages in cellulose. McGhee et al. studied the cultivation of *A. awamori* on wheat bran to produce α-galactosidase and invertase [26]. The proliferation of two different *Aspergillus* strains, including *A. awamori*, on several agro-industrial by-products, were evaluated for the production of α-galactosidase [27]. Tsakona et al. [10,11] employed the same process to formulate fermentation supplements from FRW. In that case, the authors reported on the enzymatic hydrolysis of starch to glucose. Hence, the results obtained from our study are in accordance with previous studies, indicating that the crude enzymatic extracts produced after SSF contain the essential hydrolytic enzymes to break down polysaccharides and oligosaccharides present in MFI and MCWS to generate a rich supplement for bioconversion processes (Table 1).

Therefore, the novelty of the present work focuses on the enzymatic hydrolysis of mixed confectionery waste streams containing all sugars to generate the respective monomers. This will allow for the development of a more consolidated process to exploit all waste streams, under the frame of developing bioconversion processes that can integrate and fit into the circular economy concept.

Table 1. Degree of hydrolysis (%, w/w) of starch, sucrose, and lactose during hydrolysis experiments, using varying initial solid concentrations of mixed food for infants (MFI), mixed confectionery waste streams (MCWS), and mixed waste streams (MWS) hydrolysates.

Waste Stream Concentration	Composition	MFI	MCWS	MWS
50 g·L^{-1}	Starch	95.6 ± 0.42	96.9 ± 0.89	93.6 ± 0.89
	Sucrose	91.9 ± 1.48	90.9 ± 1.96	92.1 ± 0.67
	Lactose	89.2 ± 0.92	72.3 ± 2.75	88.9 ± 1.28
100 g·L^{-1}	Starch	93.4 ± 1.06	94.5 ± 0.85	91.3 ± 1.62
	Sucrose	83.9 ± 2.26	88.9 ± 1.02	83.6 ± 1.34
	Lactose	81.8 ± 3.54	71.3 ± 0.98	78.1 ± 0.71
150 g·L^{-1}	Starch	91 ± 1.63	93.6 ± 0.56	91 ± 0.99
	Sucrose	79.4 ± 2.19	84.2 ± 1.93	78.4 ± 1.59
	Lactose	75.1 ± 2.83	70.1 ± 1.63	73.8 ± 2.59

3.2. Shake Flask Fermentations for Microbial Oil Production

As demonstrated in Table 1, the high degree of hydrolysis in MFI entailed the formulation of hydrolysates rich in glucose, fructose, and galactose. The similar composition was obtained in the hydrolysates of MCWS and MWS (Table 1). Thus, the next step would target the consumption of these sugars sources for the proliferation and microbial oil synthesis by the strain *R. toruloides*. Experiments were initiated by the fermentation of pure commercial sources of each sugar using shake flasks, and the results are presented in Table 2. It can be easily seen that after 140 h, substrate consumption was terminated yielding 36–43% of intracellular microbial oil. These experiments were the preliminary step to ensure that *R. toruloides* could consume glucose, sucrose, fructose, and galactose, thus the potential to utilize the hydrolysates obtained from MFI, MCWS, and MSW.

Table 2. Consumption of substrate, along with the production of total dry weight (TDW), microbial oil (MO), and intra-cellular oil content during the cultivation of *R. toruloides* using commercial glucose, sucrose, fructose, and galactose at 30 g·L^{-1}.

Substrate	Fermentation Time (h)	Consumed Substrate (g·L^{-1})	TDW (g·L^{-1})	MO (g·L^{-1})	Oil Content (%, w/w)
Glucose	141	30 ± 0.3	9.4 ± 0.1	3.7 ± 0.3	39.3 ± 2.4
Sucrose	147	30 ± 0.2	10.5 ± 0.3	4.6 ± 0.3	43.8 ± 1.5
Fructose	147	29.4 ± 0.9	9.7 ± 0.4	3.5 ± 0.1	36.1 ± 0.9
Galactose	147	29.6 ± 0.6	9.1 ± 0.1	3.6 ± 0.1	39.6 ± 1.9

Hence, the next step employed the utilization of these hydrolysates as the sole fermentation supplements for lipid synthesis. Figure 1 illustrates the consumption of total sugars along with TDW and lipid production expressed in g·L^{-1}, using the three different hydrolysates (a: MFI, b: MCWS, c: MWS). In all cases, the initial total sugar concentration was in the range of 80–100 g·L^{-1}, whereas FAN concentration ranged from 197–243 mg·L^{-1}. Lipid synthesis started approximately after 44 h of fermentation, triggered by nitrogen depletion in the medium (Figure 1). One of the targets of these experiments was to also identify the better performing substrate as a fermentation feedstock. The hydrolysates from MFI and MCWS resulted in a maximum 42.6% and 52.9% intracellular content of lipids, respectively, whereas in the case of MWS hydrolysate, the microbial oil content did not exceed 37.5%. The highest oil productivity was achieved in the case of MCWS (0.077 g·L^{-1}·h^{-1}), followed by MFI (0.069 g·L^{-1}·h^{-1}) and MWS (0.054 g·L^{-1}·h^{-1}). Interestingly, *R. toruloides* presented higher specific growth rate (0.28–0.29 h^{-1}) when MFI and MCWS were employed as substrates, as compared to MWS (0.12 h^{-1}). Notably, the consumption rate of total sugars was lower in the first hours of fermentation when MCWS hydrolysate was tested compared to MFI hydrolysate. More specifically, the consumption rate was 0.43 and 0.31 g·L^{-1}·h^{-1} in the case of MFI and MCWS, respectively, at the early phase of the fermentation (44 h). Still, in all used substrates, the obtained conversion yield of lipids to TDW was more than 0.3 (g g^{-1}).

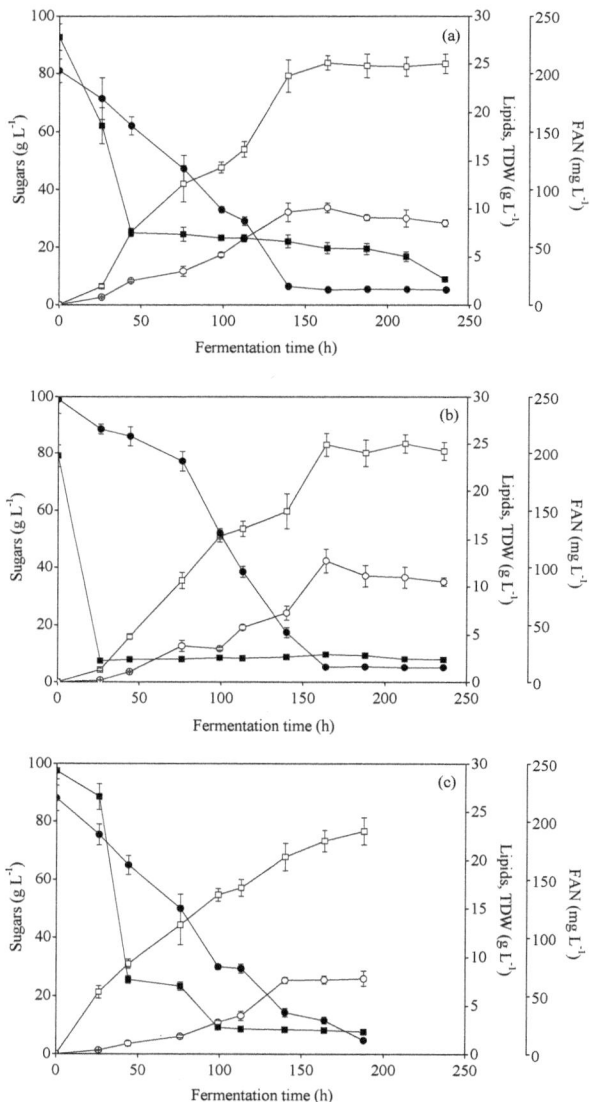

Figure 1. Concentration of sugars (●), free amino nitrogen (FAN) (■) and production of total dry weight (TDW) (□) and microbial lipids (○), during shake flask cultures of *R. toruloides* on (**a**) mixed food for infant (MFI) hydrolysates, (**b**) mixed confectionery waste streams (MCWS) hydrolysates, and (**c**) mixed waste streams (MWS) hydrolysates.

The strain *R. toruloides* constitutes an industrially important strain for the production of microbial lipids, carotenoids, and various enzymes [8]. Xu et al. reported the ample range of carbohydrates strains belonging to *Rhodotorula* species could consume, including glucose, fructose, xylose, arabinose, sucrose, starch, inulin, and glycerol [7]. In particular, for *R. toruloides*, the ability to synthesize lipids via cultivation on cassava starch, glucose, xylose, glycerol, and distillery wastewater has been well established [7].

So far, co-cultivation of *R. toruloides* was studied in the viewpoint of exploiting the hydrolysates obtained from lignocellulosic biomass, targeting either lipids or even carotenoids production [28,29]. Contradictive results have been reported based on the combination of substrates and the specificity of strains. For instance, Matsakas et al. demonstrated the parallel consumption of glucose and fructose using the strain *R. toruloides* CCT 0783 on dried sorghum stalks, achieving a final lipid concentration of 13.77 g·L^{-1} [28]. On the other hand, the combination of glucose, xylose, and arabinose resulted in the utilization of xylose after glucose depletion, whereby arabinose was not metabolized [30]. The authors employed detoxified sugarcane bagasse hydrolysates, reaching a lipid production equal to 12.3 ± 0.5. Similarly, Martins et al. evaluated a carob pulp syrup for carotenoids production during fed-batch fermentation of *R. toruloides* NCYC 921, where they demonstrated that glucose was first metabolized, whereas sucrose was not consumed, thus indicating a growth-limiting factor [29]. Bommareddy et al. presented a reconstructed metabolic model, using genomic and proteomic approach, on cell growth and synthesis of triglycerides (TAG) on glycerol, glucose xylose, arabinose, and various combinations [31]. Fermentation of glycerol and glucose presented catabolite repression phenomena whereby causing the sequential consumption of glycerol after glucose.

Table 3 shows the fatty acid profile of *R. toruloides* lipids generated during shake flask cultures on commercial sugars (glucose, sucrose, fructose, and galactose). The major fatty acids were oleic (C18:1) and palmitic (C16:0), whereas lower quantities of stearic (C18:0) and linoleic (C18:2) acids were observed. Oleic and palmitic acids corresponded to more than 75% (w/w) of the total amount of fatty acids synthesized by *R. toruloides*. This is in accordance with the results reported by Tchakouteu et al. [32], whereby oleic and palmitic acid reached 76.7% of the total fraction of lipids produced. Slininger et al. reported that cultivation of *R. toruloides* NRRL Y-1091 using lignocellulosic hydrolysate resulted in 72.9% of oleic and palmitic [33], whereas in another study, *R. toruloides* Y4 reached 66.9%, after 134 h of fed-batch fermentation [34].

Table 3. Fatty acid composition of lipids produced during shake flask cultures of *R. toruloides* on commercial sugars (glucose, sucrose, fructose, galactose).

Fermentation Time (h)	C14:0	C16:0	$^{\Delta9}$ C16:1	C18:0	$^{\Delta9}$ C18:1	$^{\Delta9,12}$ C18:2	$^{\Delta9,12,15}$ C18:3
			Glucose				
60	1.8	34	0.3	7.6	43.2	8.4	4.13
92	1.3	27.4	0.8	8	46.9	9.5	3.5
140	0.9	24.5	0.8	5.9	51.5	12.3	4
			Sucrose				
23	1.9	34.7	-	7.9	42.8	8.6	4.1
103	1.2	27.1	0.8	8.5	48.9	9.7	3.7
147	0.9	24.9	0.8	6.1	50.7	12.4	4.1
			Fructose				
45	1.6	28.1	1.1	7.5	46.8	10.9	2.7
103	1.4	27.9	0.7	7.9	48.7	9.5	3.6
147	0.9	24.5	0.9	5.7	51.5	12.3	3.7
			Galactose				
24	1.2	26.2	0.9	6.8	45.3	11.1	2.7
105	1.6	27.5	0.9	7.5	49.7	8.9	3.8
140	0.9	24.9	0.9	6.8	50.3	11.3	4.6

Similarly, in our previous study using FRW hydrolysate during fed-batch bioreactor experiments, almost 80% of the total concentration of fatty acid corresponded to oleic and palmitic acid [10].

Foods **2019**, *8*, 300

3.3. Fed-Batch Bioreactor Cultures Using MFI and MCWS Hydrolysates

One of the major targets of this study was to identify the potential of MFI, MCWS, and MWS hydrolysates as sole fermentation substrates for lipids production. Following the results from shake flask cultures, the next step employed the evaluation of the best performing substrates (MFI and MCSW) in bioreactor cultures.

Nitrogen limitation is known to induce de novo synthesis of lipids in oleaginous strains as it is a secondary activity, occurring while the carbon source is in excess [35,36]. Deficiency of nitrogen in the fermentation broth leads to the low activity of isocitrate dehydrogenase; thus, metabolic flux is directed to lipid overproduction [35,37]. This case is induced at high carbon to nitrogen (C/N) ratios, whereby biomass production is hindered, and the carbon source is channeled into lipogenesis [37].

In this study, the initial C/N ratio was selected following our previous work with *R. toruloides* and FRW hydrolysates [10], hence adjusted to ~70–80 g·g^{-1}. Figure 2a presents the time course consumption of total sugars (glucose, fructose, and galactose), along with FAN utilization, whereas in Figure 2b, the production of TDW, microbial lipids, and IPS are presented. Initial total sugar concentration was ~50 g·L^{-1}, with glucose being the main sugar, whereas fructose and galactose were present in low concentrations (5.4 and 6.4 g·L^{-1}, respectively). Intermittent additions of the feeding solution were performed, when sugars' concentration was lower than 20 g·L^{-1}. Maximal concentration of microbial lipids was observed after 92 h of fermentation (16.6 g·L^{-1}) with an intracellular content of 43.3% (w/w) and a productivity of 0.18 ·L^{-1}·h^{-1}.

Similarly, Figure 3 depicts the results obtained when MCWS hydrolysate was evaluated by applying a fed-batch strategy. In this case, lipid production reached 17 g·L^{-1} after 98 h, resulting in productivity of 0.17 g·L^{-1}·h^{-1}. The lipid content was 45.6% (w/w).

It is easily observed that when both MFI and MCWS hydrolysates were employed, FAN depletion from the media channeled carbon flux to lipid overproduction. This observation was also confirmed by the fact that RCW remained constant until the end of fermentation. It is also interesting to note that consumption of glucose, fructose, and galactose occurred at the same time in both cases, without indicating carbon catabolite repression phenomena in the presence of glucose. These results are in agreement with the results reported by Matsakas et al., where glucose and fructose were equally consumed [28].

Studies on the simultaneous consumption of carbon sources have been recently initiated, and significant work is undertaken on transcriptomic and proteomic levels to understand the metabolic network [31,38]. Bommareddy et al. presented a metabolic model of *R. toruloides* by using different carbon sources. When glucose was compared with glycerol during bioreactor fermentations, increased biomass was noted in the case of glucose, regardless of the final production of lipids, which was almost equal in both cases [31]. Following flux distribution analysis, the authors showed that when glucose comprised the sole carbon and energy source, 63% of NADPH derived from the pentose phosphate (PP) pathway, complimented by the cytosolic malic enzyme [31]. On the other hand, when glycerol alone was employed, NADPH was supplied by the cytosolic malic enzyme, indicating a reduction in PP pathway. Similarly, using the pentoses xylose and arabinose, high PP pathway was observed to meet the demand for NADPH uptake [31]. The same authors, further performed transcriptomics to investigate gene expression during the combined fermentation of glucose with glycerol [38]. Significant findings concerning the upregulation and downregulation of genes were presented, particularly during the phase of nitrogen starvation [38].

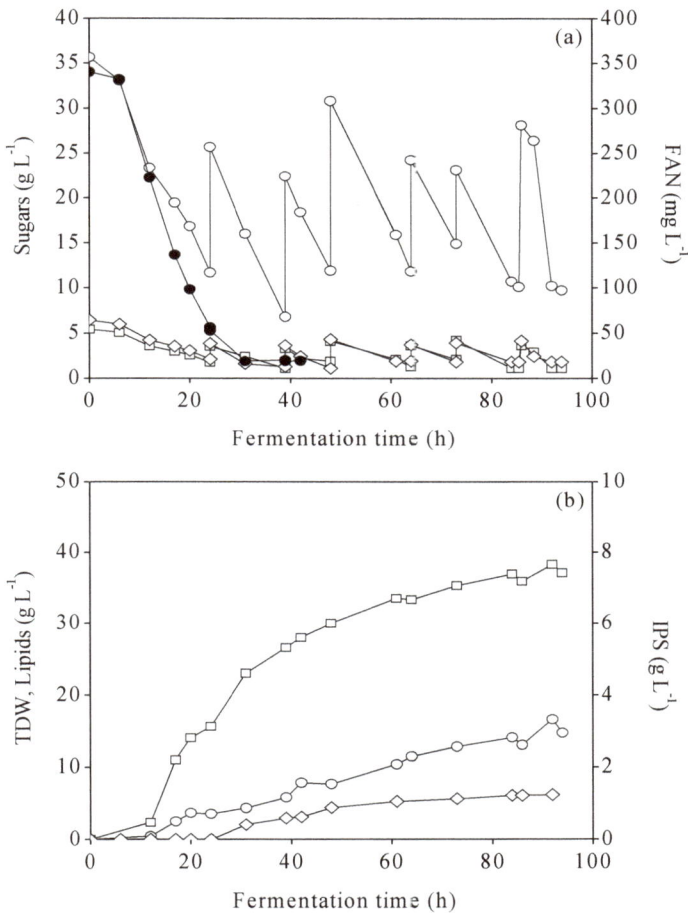

Figure 2. (**a**) Concentration of glucose (○), fructose (□), galactose (◇), free amino nitrogen (FAN) (●) and (**b**) production of total dry weight (TDW) (□), microbial lipids (○), intracellular polysaccharides (IPS) (◇) during fed-batch bioreactor fermentations of *R. toruloides* on mixed food for infants (MFI) hydrolysates.

Combination of glucose with glycerol entailed a diauxic growth; however, glycerol addition improved lipid synthesis. Wiebe et al. evaluated C5 and C6 sugars, particularly glucose, xylose, arabinose, and their combinations, during batch and fed-batch cultures. They stated that lipid production in the mixture was lower compared to individual application of glucose or xylose, regardless of the proportions of glucose and xylose in the mixture [39]. In another study, Easterling et al. used the oleaginous strain *R. glutinis* during fermentation of dextrose, xylose, and glycerol and their mixtures, demonstrating that mix substrate cultivations resulted in increased biomass production compared to individual carbon sources [40]. On top of that, the strain *R. toruloides* has been previously shown to produce IPS [11,32], which was also observed in this study (~1.2 g·L^{-1}). Analysis of IPS by HPLC at the end of the fermentation demonstrated that they were primarily (>60%) comprised of mannose and glucose and to a lesser extent of galactose and fructose.

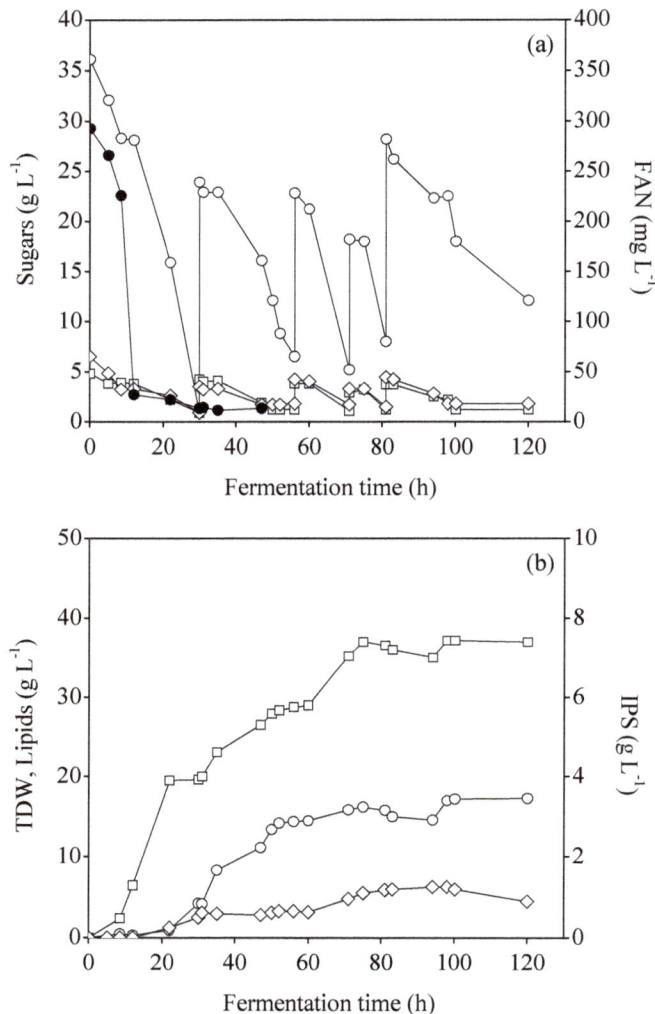

Figure 3. (**a**) Concentration of glucose (○), fructose (□), galactose (◊), free amino nitrogen (FAN) (●) and (**b**) production of total dry weight (TDW) (□), microbial lipids (○), intracellular polysaccharides (IPS) (◊) during fed-batch bioreactor fermentations of *R. toruloides* on mixed confectionery waste streams (MCWS) hydrolysates.

Table 4 presents the fatty acid profile of the microbial lipids produced during fed-batch experiments using MFI and MCWS hydrolysates. As expected, oleic (C18:1), stearic (C18:0), palmitic (C16:0), and linoleic (C18:2) were the major fatty acids identified. Compared to FRW employed in our previous work [10], the proportions of the individual fatty acid were modified, an observation deriving probably by the different composition of the substrate. More specifically, oleic acid (C18:1) remained the major fatty acid produced; however, it was increased from approximately 52.5–61.4% when MFI and MCWS were used. When the latter substrates were employed, the fractions of palmitic acid (C16:0) and linoleic (C18:2) were significantly reduced, whereby the fractions of stearic acid (C18:0) increased almost 2-fold compared to FRW hydrolysates. Similarly, the fraction of palmitic acid was decreased when MFI and

MCWS were used in fed-batch experiments compared to shake flask cultures, where individual carbon sources were used. On the other hand, oleic acid was increased when MFI and MCWS were applied for lipid bioconversion. The differences found in fatty acid composition when different substrates (FRW, MFI, and MCWS) were employed may be attributed to their different sugar composition. In fact, FRW contains only glucose, whereas MFI and MCWS contain also fructose and galactose. The effect of substrate composition on the fatty acid profile of *R. toruloides* microbial oil has been highlighted also by previous studies [7,14]. Further research studies focused on proteomics and transcriptomics have demonstrated that gene expression and substrate specificity have a key role in lipid synthesis [41]. Specifically, Fillet et al. mentioned that the final chain length of the fatty acid relates to the substrate preference of the elongase 3-ketoacyl-CoA [42]. Furthermore, Zhu et al. pointed that the yield of polyunsaturated fatty acids depends on the number of an acyl carrier protein (ACP), which is part of the fatty acid synthase system, exhibiting a significant role in the chain-elongation process [36].

Table 4. Fatty acid composition of plant-derived oils compared with the composition of the microbial lipids produced during fed-batch bioreactor cultures of *R. toruloides* using the hydrolysates from mixed food for infants (MFI) and mixed confectionery waste streams (MCWS).

Oil Source	Fatty Acids (%)							Reference
	C14:0	C16:0	$^{\Delta 9}$ C16:1	C18:0	$^{\Delta 9}$ C18:1	$^{\Delta 9,12}$ C18:2	$^{\Delta 9,12,15}$ C18:3	
Soybean	–	6–10	0.1	2–5	20–24.9	50–60	4.3–11	[43–47]
Rapeseed	–	2.8–14		0.9–2	13.6–64.1	11.8–26	7.5–13.2	[43–47]
Cottonseed	–	27–28.7	–	0.9–2	13–18	51–58	8	[43–47]
Sunflower	–	4.6–6.4	0.1	2.9–3.7	17–62.8	27.5–74	0.1–0.2	[43–47]
Palm oil	0.7	36.7–44	0.1	5–6.6	3–46.1	8.6–11	0.3	[43–47]
Olive oil	<0.1	7.5–20	0.3–3.5	0.5–5	55–83	3.5–21.0	≤1.0	[48]
Microbial lipids from bioreactor cultures								
FRW [a]	1.5	28.7	0.6	7.5	50.3	9.5	1.4	[10]
MFI	1.4	10.3	0.7	14.5	61.2	5.3	0.4	This study
MCWS	0.9	15.2	0.9	13.8	61.7	6.1	0.1	

[a] FRW: flour-rich waste.

The applied feeding strategy during fed-batch cultures, oxygen saturation, and primarily carbon source constitute key factors on the final profile analysis of TAGs produced. The effect of feeding strategy and dissolved oxygen was beyond the scope of the current study; thus, the emphasis was given on the carbon source. When FRW hydrolysates were used [10], unsaturated content was found to be 61.8%, a proportion that increased in both cases where MFI and MCWS were employed (72.1 and 69.7%, respectively). It is generally accepted that de novo synthesis of lipids in *R. toruloides* results mainly in unsaturated fatty acids (e.g., oleic and linoleic acid) [7]. Fei et al. used corn stover hydrolysates and reported that oleic acid (C18:1) was the major fatty acid produced in all applied strategies, followed by palmitic (C16:0) [49]. Wiebe et al. reported that stearic, linoleic, and palmitic fractions were affected in the presence of mixed carbon sources compared to pure glucose fermentation experiments, whereby xylose and arabinose addition induced an increase in C16:0 and C18:2 fatty acids [39]. Patel et al. noted that palmitic acid was not detected when glucose and fructose were used as fermentation sources with *R. kratochvilovae* HIMPA1 [50]. The presence of glucose and fructose led to increased monounsaturated fatty acids, whereas sucrose increased polyunsaturated fatty acids. Interestingly, Bommareddy et al. [31] showed a content of 57% in saturated fatty acids when glucose alone was employed, whereas, during shake flask cultures in our study, the corresponding amount reached approximately 37.6% (Table 3). This could be attributed to the conditions during fermentation (bioreactor compared to shaking flasks), thus indicating future research. Zeng et al. undertook the utilization of food waste hydrolysates containing glucose as the major carbon source during flask experiments, demonstrating that oleic (75.8%) was the predominant fatty acid, followed by palmitic, linoleic, and lower quantities of stearic [51]. As previously noted, oxygen in the fermentation medium is also crucial for the formulation of lipid fractions. Minkevich et al. also stated the variation between saturated and unsaturated fatty acid profile with oxygen limitation [52], whereby Bommareddy et al.

pointed the production of saturated fatty acids following un-controlled oxygen supply [31]. On the other hand, the provision of glycerol in the fermentation media resulted in increased saturated fractions. For instance, a combination of crude glycerol and sunflower meal hydrolysates [9] resulted in 43.1% of saturated fatty acids, whereas Signori et al. reported 42.2–42.4% of saturated content when pure and crude glycerol were used, respectively [53].

Previous studies on microbial oil production through bioconversion processes focus primarily to evaluate the lipids for biodiesel production through transesterification processes [54,55]. However, under the viewpoint of bio-economy transition, emphasis should be also given to novel applications of microbial lipids for the production of lipid-based products with improved quality and specific applications [13]. The production of food products with a limited amount of saturated fat has emerged to be of paramount importance for food industries during the last years. For instance, high unsaturated vegetable oils (i.e., soybean oil) are preferred for the production of oleogels, which are considered a healthier substitute for trans and saturated fats in food products [17]. Bharathiraja et al. reviewed the application of microbial lipids in food formulations, stating how they could replace plant-derived lipids (e.g., cocoa butter, palm oil) and further utilized as stabilizing and thickening agents, emulsifiers, and water-binding compounds [56]. Equally, given the high content of oleic acid in the produced lipids, it could be also used as a substitute for cocoa butter, based on the high content of unsaturated fatty acids.

Apart from texture, fatty acids can also regulate the aroma and flavor of specific types of food. Likewise, fatty acids can be implemented in pharmaceutical applications, and more specifically, to formulate fortified foods and/or beverages with polyunsaturated fatty acids, entailing possible health effects [56]. Concerning microbial lipids from *R. toruloides*, Papadaki et al. conducted a study on the valorization of yeast lipid derivatives to formulate bio-based wax esters through a two-stage biocatalysis reaction [57]. Implementation of molasses as fermentation supplement led to the production of 8.1 g·L^{-1} microbial oil, containing mainly oleic acid (51%), palmitic acid, and stearic acid. The generated oleogels, using olive oil as the base oil, successfully simulated commercial margarine, thus demonstrating their potential for future applications in spreadable fat-products [57]. It should be stressed that the results obtained in the present study highlight that the valorization of the whole confectionery waste streams potential through the proposed process enhance the feasibility for further integration in existing facilities. Likewise, the proposed scheme leads to the production of lipids with a higher degree of unsaturation, a result that could be exploited under the frame of the circular economy towards the development of high value-added products for targeted food formulations within the initial industry.

4. Conclusions

The potential of generating nutrient-rich fermentation supplements deriving from diversified confectionery waste streams has been well established. All the evaluated hydrolysates performed well both in shake flask cultures and fed-batch bioreactor cultivations using *R. toruloides* to produce microbial lipids. Implementation of mixed substrates containing glucose, fructose, and galactose did not affect the consumption rates, however, resulted in modifications in the fractions of fatty acids. The possibility to produce tailor-made microbial lipid fractions for the production of lipid-based products was also presented. A consolidated bioprocess previously developed to valorize mixed confectionery waste streams could be further expanded to also target future food applications of microbial lipids under the context of the bio-economy era.

Author Contributions: Conceptualization, N.K., S.P., and A.K.; Methodology, S.T., A.P., V.K., and N.K.; Investigation, S.T. and A.P.; Resources, S.T., A.P., and V.K.; Writing—Original Draft Preparation, S.T., A.P., and V.K.; Writing—Review and Editing, N.K., S.P., and A.K.; Supervision, N.K. and A.K.

Funding: This work is funded by the research project "NUTRI-FUEL" (09SYN-32-621), implemented within the National Strategic Reference Framework (NSRF) 2007–2013 and co-financed by National (Greek Ministry—General Secretariat of Research and Technology) and Community Funds (E.U.—European Social Fund).

Conflicts of Interest: The authors declare no conflict of interest.

Abbreviations: FRW—Flour-rich waste streams, SSF—Solid-state fermentation, MFI—Mixed food for infants, MCWS—Mixed confectionery waste streams, MWS—Mixed waste streams, FAN—Free amino nitrogen, TDW—Total dry weight, RCW—Residual cell weight, IP—Inorganic phosphorus, MO—Microbial oil, FAME—Fatty acid methyl esters, IPS—Intracellular polysaccharides, WMB—Wheat milling by-products, TAG—Triglycerides.

References

1. Ghisellini, P.; Cialani, C.; Ulgiati, S. A review on circular economy: The expected transition to a balanced interplay of environmental and economic systems. *J. Clean Prod.* **2016**, *114*, 11–32. [CrossRef]
2. Maina, S.; Kachrimanidou, V.; Koutinas, A. A roadmap towards a circular and sustainable bioeconomy through waste valorization. *Curr. Opin. Green Sustain. Chem.* **2017**, *8*, 18–23. [CrossRef]
3. Ochsenreither, K.; Glück, C.; Stressler, T.; Fischer, L.; Syldatk, C. Production Strategies and Applications of Microbial Single Cell Oils. *Front. Microbiol.* **2016**, *7*, 1539. [CrossRef] [PubMed]
4. Athenaki, M.; Gardeli, C.; Diamantopoulou, P.; Tchakouteu, S.S.; Sarris, D.; Philippoussis, A.; Papanikolaou, S. Lipids from yeasts and fungi: Physiology, production and analytical considerations. *J. Appl. Microbiol.* **2018**, *124*, 336–367. [CrossRef] [PubMed]
5. Koutinas, A.A.; Chatzifragkou, A.; Kopsahelis, N.; Papanikolaou, S.; Kookos, I.K. Design and techno-economic evaluation of microbial oil production as a renewable resource for biodiesel and oleochemical production. *Fuel* **2014**, *116*, 566–577. [CrossRef]
6. Kopsahelis, N.; Dimou, C.; Papadaki, A.; Xenopoulos, E.; Kyraleou, M.; Kallithraka, S.; Kotseridis, G.; Papanikolaou, S.; Kookos, I.K.; Koutinas, A.A. Refining of wine lees and cheese whey for the production of microbial oil, polyphenol-rich extracts and value-added co-products. *J. Chem. Techol. Biotechnol.* **2018**, *93*, 257–268. [CrossRef]
7. Xu, J.; Liu, D. Exploitation of genus *Rhodosporidium* for microbial lipid production. *World J. Microbiol. Biotechnol.* **2017**, *33*, 54. [CrossRef] [PubMed]
8. Park, Y.K.; Nicaud, J.M.; Ledesma-Amaro, R. The Engineering Potential of *Rhodosporidium toruloides* as a Workhorse for Biotechnological Applications. *Trends Biotechnol.* **2018**, *36*, 304–317. [CrossRef]
9. Leiva-Candia, D.E.; Tsakona, S.; Kopsahelis, N.; García, I.L.; Papanikolaou, S.; Dorado, M.P.; Koutinas, A.A. Biorefining of by-product streams from sunflower-based biodiesel production plants for integrated synthesis of microbial oil and value-added co-products. *Bioresour. Technol.* **2015**, *190*, 57–65. [CrossRef]
10. Tsakona, S.; Skiadaresis, A.G.; Kopsahelis, N.; Chatzifragkou, A.; Papanikolaou, S.; Kookos, I.K.; Koutinas, A.A. Valorisation of side streams from wheat milling and confectionery industries for consolidated production and extraction of microbial lipids. *Food Chem.* **2016**, *198*, 85–92. [CrossRef]
11. Tsakona, S.; Kopsahelis, N.; Chatzifragkou, A.; Papanikolaou, S.; Kookos, I.K.; Koutinas, A.A. Formulation of fermentation media from flour-rich waste streams for microbial lipid production by *Lipomyces starkeyi*. *J. Biotechnol.* **2014**, *189*, 36–45. [CrossRef] [PubMed]
12. Papadaki, A.; Fernandes, K.V.; Chatzifragkou, A.; Aguieiras, E.C.G.; da Silva, J.A.C.; Fernandez-Lafuente, R.; Papanikolaou, S.; Koutinas, A.; Freire, D.M.G. Bioprocess development for biolubricant production using microbial oil derived via fermentation from confectionery industry waste. *Bioresour. Technol.* **2018**, *267*, 311–318. [CrossRef] [PubMed]
13. Papadaki, A.; Mallouchos, A.; Efthymiou, M.-N.; Gardeli, C.; Kopsahelis, N.; Aguieiras, E.C.G.; Freire, D.M.G.; Papanikolaou, S.; Koutinas, A.A. Production of wax esters via microbial oil synthesis from food industry waste and by-product streams. *Bioresour. Technol.* **2017**, *245*, 274–282. [CrossRef] [PubMed]
14. Boviatsi, E.; Papadaki, A.; Efthymiou, M.-N.; Nychas, G.-J.E.; Papanikolaou, S.; da Silva, J.A.C.; Freire, D.M.G.; Koutinas, A. Valorisation of sugarcane molasses for the production of microbial lipids via fermentation of two *Rhodosporidium* strains for enzymatic synthesis of polyol esters. *J. Chem. Technol. Biotechnol.* **2019**. [CrossRef]
15. Wu, H.; Li, Y.; Chen, L.; Zong, M. Production of microbial oil with high oleic acid content by *Trichosporon capitatum*. *Appl. Energ.* **2011**, *88*, 138–142. [CrossRef]

16. Pulidindi, K.; Chakraborty, S. *Tall Oil Fatty Acid Market Size By Product (Oleic Acid, Linoleic Acid), By Application (Dimer Acids, Alkyd Resins, Fatty Acid Esters), By End-Use (Soap &Detergents, Coatings, Lubricants, Plastics, Fuel Additives, Metal Working Fluid), Industry Analysis Report, Regional Outlook (U.S., Canada, Germany, UK, France, Spain, Italy, China, India, Japan, Australia, Indonesia, Malaysia, Brazil, Mexico, South Africa, GCC), Application Growth Potential, Price Trends, Competitive Market Share & Forecast, 2017–2024*; Global Market Insights: Selbyville, DE, USA, 2017.

17. Mert, B.; Demirkesen, I. Evaluation of highly unsaturated oleogels as shortening replacer in a short dough product. *LWT - Food Sci. Technol.* **2016**, *68*, 477–484. [CrossRef]

18. Food and Drug Administration. Final determination regarding partially hydrogenated oils. Notification; declaratory order; extension of compliance date. *Fed. Regist.* **2018**, *83*, 23358–23359.

19. Folch, J.; Lees, M.; Sloane-Stanley, G.H. A simple method for the isolation and purification of total lipids from animal tissues. *J. Biol. Chem.* **1957**, *226*, 497–509. [PubMed]

20. Liang, Y.; Sarkany, N.; Cui, Y.; Blackburn, J.W. Batch stage study of lipid production from crude glycerol derived from yellow grease or animal fats through microalgal fermentation. *Bioresour. Technol.* **2010**, *101*, 6745–6750. [CrossRef]

21. Blandino, A.; Iqbalsyah, T.; Pandiella, S.; Cantero, D.; Webb, C. Polygalacturonase production by *Aspergillus awamori* on wheat in solid-state fermentation. *Appl. Microbiol. Biotechnol.* **2002**, *58*, 164–169. [CrossRef]

22. Díaz, A.B.; Alvarado, O.; de Ory, I.; Caro, I.; Blandino, A. Valorization of grape pomace and orange peels: Improved production of hydrolytic enzymes for the clarification of orange juice. *Food Bioprod. Process.* **2013**, *91*, 580–586. [CrossRef]

23. Smaali, I.; Soussi, A.; Bouallagui, H.; Chaira, N.; Hamdi, M.; Marzouki, M.N. Production of high-fructose syrup from date by-products in a packed bed bioreactor using a novel thermostable invertase from *Aspergillus awamori*. *Biocatal. Biotransformation* **2011**, *29*, 253–261. [CrossRef]

24. Bertolin, T.E.; Schmidell, W.; Maiorano, A.E.; Casara, J.; Costa, J.A.V. Influence of Carbon, Nitrogen and Phosphorous Sources on Glucoamylase Production by *Aspergillus awamori* in Solid State Fermentation. *Z. Naturforsch. C* **2003**, *58*, 708–712. [CrossRef] [PubMed]

25. Botella, C.; De Ory, I.; Webb, C.; Cantero, D.; Blandino, A. Hydrolytic enzyme production by *Aspergillus awamori* on grape pomace. *Biochem. Eng. J.* **2005**, *26*, 100–106. [CrossRef]

26. McGhee, J.E.; Silmaim, R.; Bagley, E.B. Production of α-galactosidase from *Aspergillus awamori*: Properties and action on para-nitrophenyl α-D-galactopyranoside and galacto-Oligosaccharides of soy milk. *J. Am. Oil Chem. Soc.* **1978**, *55*, 244–247. [CrossRef]

27. Ali, U.F.; El-Gindy, A.A.; Ali, U.F.; Ibrahim, Z.M.; Isaac, G.S. A Cost-effective Medium for Enhanced Production of Extracellular α-galactosidase in Solid Substrate Cultures of *Aspergillus awamori* and *A. carbonarius*. *Aust. J. Basic Appl. Sci.* **2008**, *2*, 880–899.

28. Matsakas, L.; Bonturi, N.; Miranda, E.; Rova, U.; Christakopoulos, P. High concentrations of dried sorghum stalks as a biomass feedstock for single cell oil production by *Rhodosporidium toruloides*. *Biotechnol. Biofuels* **2015**, *8*, 6. [CrossRef]

29. Martins, V.; Dias, C.; Caldeira, J.; Duarte, L.C.; Reis, A.; Lopes da Silva, T. Carob pulp syrup: A potential Mediterranean carbon source for carotenoids production by *Rhodosporidium toruloides* NCYC 921. *Bioresour. Technol. Reports* **2018**, *3*, 177–184. [CrossRef]

30. Zhao, X.; Peng, F.; Du, W.; Liu, C.; Liu, D. Effects of some inhibitors on the growth and lipid accumulation of oleaginous yeast *Rhodosporidium toruloides*. *Bioprocess. Biosyst. Eng.* **2012**, *35*, 993–1004. [CrossRef]

31. Bommareddy, R.R.; Sabra, W.; Maheshwari, G.; Zeng, A.P. Metabolic network analysis and experimental study of lipid production in *Rhodosporidium toruloides* grown on single and mixed substrates. *Microb. Cell. Fact.* **2015**, *14*. [CrossRef]

32. Tchakouteu, S.S.; Kopsahelis, N.; Chatzifragkou, A.; Kalantzi, O.; Stoforos, N.G.; Koutinas, A.A.; Aggelis, G.; Papanikolaou, S. *Rhodosporidium toruloides* cultivated in NaCl-enriched glucose-based media: Adaptation dynamics and lipid production. *Eng. Life Sci.* **2016**, *17*, 237–248. [CrossRef]

33. Slininger, P.J.; Dien, B.S.; Kurtzman, C.P.; Moser, B.R.; Bakota, E.L.; Thompson, S.R.; O'Bryan, P.J.; Cotta, M.A.; Balan, V.; Jin, M.; et al. Comparative Lipid Production by Oleaginous Yeasts in Hydrolyzates of Lignocellulosic Biomass and Process Strategy for High Titers. *Biotechnol. Bioeng.* **2016**, *113*, 1676–1690. [CrossRef] [PubMed]

34. Li, Y.; Zhao, Z.; Bai, F. High-density cultivation of oleaginous yeast *Rhodosporidium toruloides* Y4 in fed-batch culture. *Enzyme Microb. Technol.* **2007**, *41*, 312–317. [CrossRef]

35. Liu, H.; Zhao, X.; Wang, F.; Li, Y.; Jiang, X.; Ye, M.; Zhao, Z.K.; Zou, H. Comparative proteomic analysis of *Rhodosporidium toruloides* during lipid accumulation. *Yeast* **2009**, *26*, 553–566. [CrossRef] [PubMed]

36. Zhu, Z.; Zhang, S.; Liu, H.; Shen, H.; Lin, X.; Yang, F.; Zhou, Y.J.; Jin, G.; Ye, M.; Zou, H.; et al. A multi-omic map of the lipid-producing yeast *Rhodosporidium toruloides*. *Nat. Commun.* **2012**, *3*, 1112. [CrossRef] [PubMed]

37. Castañeda, M.T.; Nuñez, S.; Garelli, F.; Voget, C.; De Battista, H. Comprehensive analysis of a metabolic model for lipid production in *Rhodosporidium toruloides*. *J. Biotechnol.* **2018**, *280*, 11–18. [CrossRef] [PubMed]

38. Bommareddy, R.R.; Sabra, W.; Zeng, A.P. Glucose-mediated regulation of glycerol uptake in *Rhodosporidium toruloides*: Insights through transcriptomic analysis on dual substrate fermentation. *Eng. Life Sci.* **2017**, *17*, 282–291. [CrossRef]

39. Wiebe, M.G.; Koivuranta, K.; Penttilä, M.; Ruohonen, L. Lipid production in batch and fed-batch cultures of *Rhodosporidium toruloides* from 5 and 6 carbon carbohydrates. *BMC Biotechnol.* **2012**, *12*. [CrossRef]

40. Easterling, E.R.; French, W.T.; Hernandez, R.; Licha, M. The effect of glycerol as a sole and secondary substrate on the growth and fatty acid composition of *Rhodotorula glutinis*. *Bioresour. Technol.* **2009**, *100*, 356–361. [CrossRef]

41. Chen, Z.; Liu, P.; Liu, Y.; Tang, H.; Chen, Y.; Zhang, L. Identification and characterization of a type-2 diacylglycerol acyltransferase (DGAT2) from *Rhodosporidium diobovatum*. *Antonie Van Leeuwenhoek* **2014**, *106*, 1127–1137. [CrossRef]

42. Fillet, S.; Ronchel, C.; Callejo, C.; Fajardo, M.-J.; Moralejo, H.; Adrio, J.L. Engineering *Rhodosporidium toruloides* for the production of very long-chain monounsaturated fatty acid-rich oils. *Appl. Microbiol. Biotechnol.* **2017**, *101*, 7271–7280. [CrossRef] [PubMed]

43. Van Gerpen, J.; Shanks, B.; Pruszko, R.; Clements, D.; Knothe, G. *Biodiesel Production Technology: August 2002–January 2004*; National Renewable Energy Laboratory: Golden, CO, USA, 2004.

44. Ratledge, C. Single cell oils—have they a biotechnological future? *Trends Biotechnol.* **1993**, *11*, 278–284. [CrossRef]

45. Singh, S.P.; Singh, D. Biodiesel production through the use of different sources and characterization of oils and their esters as the substitute of diesel: A review. *Renew. Sust. Energ. Rev.* **2010**, *14*, 200–216. [CrossRef]

46. Thompson, J.C.; He, B.B. Characterization of crude glycerol from biodiesel production from multiple feedstocks. *Appl. Eng. Agric.* **2006**, *22*, 261–265. [CrossRef]

47. Ramos, M.J.; Fernández, C.M.; Casas, A.; Rodríguez, L.; Pérez, Á. Influence of fatty acid composition of raw materials on biodiesel properties. *Bioresour. Technol.* **2009**, *100*, 261–268. [CrossRef] [PubMed]

48. Boskou, D.; Blekas, G.; Tsimidou, M. Olive oil composition. In *Olive Oil*, 2nd ed.; Boskou, D., Ed.; Elsevier Inc. AOCS Press: Champaign, IL, USA, 2006; pp. 41–72. [CrossRef]

49. Fei, Q.; O'Brien, M.; Nelson, R.; Chen, X.; Lowell, A.; Dowe, N. Enhanced lipid production by *Rhodosporidium toruloides* using different fed-batch feeding strategies with lignocellulosic hydrolysate as the sole carbon source. *Biotechnol. Biofuel.* **2016**, *9*. [CrossRef]

50. Patel, A.; Pruthi, V.; Singh, R.P.; Pruthi, P.A. Synergistic effect of fermentable and non-fermentable carbon sources enhances TAG accumulation in oleaginous yeast *Rhodosporidium kratochvilovae* HIMPA1. *Bioresour. Technol.* **2015**, *188*, 136–144. [CrossRef]

51. Zeng, Y.; Xie, T.; Li, P.; Jian, B.; Li, X.; Xie, Y.; Zhang, Y. Enhanced lipid production and nutrient utilization of food waste hydrolysate by mixed culture of oleaginous yeast *Rhodosporidium toruloides* and oleaginous microalgae *Chlorella vulgaris*. *Renew. Energ.* **2018**, *126*, 915–923. [CrossRef]

52. Minkevich, I.; Dedyukhina, E.G.; Chistyakova, T.I. The effect of lipid content on the elemental composition and energy capacity of yeast biomass. *Appl. Microbiol. Biotechnol.* **2010**, *88*, 799–806. [CrossRef]

53. Signori, L.; Ami, D.; Posteri, R.; Giuzzi, A.; Mereghetti, F.; Porro, D.; Branduardi, P. Assessing an effective feeding strategy to optimize crude glycerol utilization as sustainable carbon source for lipid accumulation in oleaginous yeasts. *Microb. Cell Factor.* **2016**, *15*, 75. [CrossRef]

54. Leiva-Candia, D.E.; Pinzi, S.; Redel-Macías, M.D.; Koutinas, A.; Webb, C.; Dorado, M.P. The potential for agro-industrial waste utilization using oleaginous yeast for the production of biodiesel. *Fuel* **2014**, *123*, 33–42. [CrossRef]

55. Cheirsilp, B.; Louhasakul, Y. Industrial wastes as a promising renewable source for production of microbial lipid and direct transesterification of the lipid into biodiesel. *Bioresour Technol.* **2013**, *142*, 329–337. [CrossRef] [PubMed]

56. Bharathiraja, B.; Sridharan, S.; Sowmya, V.; Yuvaraj, D.; Praveenkumar, R. Microbial oil—A plausible alternate resource for food and fuel application. *Bioresour. Technol.* **2017**, *233*, 423–432. [CrossRef] [PubMed]
57. Papadaki, A.; Cipolatti, E.P.; Aguieiras, E.C.G.; Pinto, M.C.C.; Kopsahelis, N.; Freire, D.M.G.; Mandala, I.; Koutinas, A.A. Development of Microbial Oil Wax-Based Oleogel with Potential Application in Food Formulations. *Food Bioprocess. Technol.* **2019**, *12*, 899–909. [CrossRef]

![foods logo] *foods*

MDPI

Article

The Development of a Sorghum Bran-Based Biorefining Process to Convert Sorghum Bran into Value Added Products

Oyenike Makanjuola [1], Darren Greetham [1], Xiaoyan Zou [1,2] and Chenyu Du [1,*]

[1] School of Applied Sciences, University of Huddersfield, Queensgate, Huddersfield HD1 3DH, UK
[2] Key Laboratory of Functional Inorganic Material Chemistry, Heilongjiang University, Harbin 150080, China
*   Correspondence: c.du@hud.ac.uk

Received: 16 June 2019; Accepted: 22 July 2019; Published: 24 July 2019

Abstract: Sorghum bran, a starch rich food processing waste, was investigated for the production of glucoamylase in submerged fungal fermentation using *Aspergillus awamori*. The fermentation parameters, such as cultivation time, substrate concentration, pH, temperature, nitrogen source, mineral source and the medium loading ratio were investigated. The glucoamylase activity was improved from 1.90 U/mL in an initial test, to 19.3 U/mL at 10% (w/v) substrate concentration, pH 6.0, medium loading ratio of 200 mL in 500 mL shaking flask, with the addition of 2.5 g/L yeast extract and essential minerals. Fermentation using 2 L bioreactors under the optimum conditions resulted in a glucoamylase activity of 23.5 U/mL at 72 h, while further increase in sorghum bran concentration to 12.5% (w/v) gave an improved gluco-amylase activity of 37.6 U/mL at 115 h. The crude glucoamylase solution was used for the enzymatic hydrolysis of the sorghum bran. A sorghum bran hydrolysis carried out at 200 rpm, 55 °C for 48 h at a substrate loading ratio of 80 g/L resulted in 11.7 g/L glucose, similar to the results obtained using commercial glucoamylase. Large-scale sorghum bran hydrolysis in 2 L bioreactors using crude glucoamylase solution resulted in a glucose concentration of 38.7 g/L from 200 g/L sorghum bran, corresponding to 94.1% of the theoretical hydrolysis yield.

Keywords: glucoamylase; sorghum milling waste; submerged fungal fermentation; *Aspergillus awamori*; hydrolysis; waste valorization

1. Introduction

The increasing concerns about global energy shortages and environmental pollution have encouraged research on the development of biorefining strategies for the conversion of renewable raw materials into value added products. Various crops, such as wheat, corn and rapeseed, that are historically only used for food, however, have now been targeted as starting materials for the production of biofuels, biochemical and biopolymers. Sorghum (*Sorghum bicolor*) is a cereal plant of the grass family *Gramineae* which originates from Africa [1]. It is the 5th most important crop cereal in the world in terms of its acreage and production [2]. Sorghum is particular important to African countries, as it grows well in hot and arid regions [3]. Traditionally, sorghum grains are cleaned, conditioned, tempered and debranned in grain hulliers to remove the outermost fibrous layer. It is then milled and sieved (dry milling process) to obtain a flour fraction using a Buhler mill [4].

Besides direct human food application, sorghum crops have been used in many fields, such as animal feed, and the production of biofuels, enzymes and bioactive compounds. Figure 1 shows a schematic diagram of sorghum-based biorefinery approaches for the production of various non-food products. Biofuel is one of the fields in which sorghum has received significant attention. Li et al. reported a demonstration study of converting sweet sorghum stems into bioethanol [5]. Sixteen tons of sweet sorghum stems were used and 1 ton of ethanol (99.5% v/v) was obtained. The cost of sorghum-derived fuel bioethanol was estimated to be \$0.49 per litre [5]. It was estimated that the sweet sorghum production in China using only marginal land could reach 13.57 million tons [6]. Approximately 0.85 million tons of bioethanol could be produced [6]. Ahmed El-Iman et al. estimated the bioethanol production potential in Nigeria using sorghum bran [7], reaching the conclusion that 497 million US gallons of bioethanol could be produced, which equivalent to 17% of Nigeria's transportation fuel use. Beside bioethanol, sorghum has also been investigated for biobutanol [8], biogas [9] and biohydrogen [10] production. As for high value markets, sorghum has been reported as the basis for the synthesis of astaxanthin [11], 3-deoxyanthocyanidins and various other bioactive compounds [12]. Lolasi et al. 2018 cultivated a halotolerant bacterium *Nesterenkonia sp* on sorghum bagasse hydrolysate and obtained an α-amylase solution of 97 U/mL [13], demonstrating the feasibility of using sorghum waste streams to produce enzymes. These studies have suggested that sorghum and sorghum processing waste have a huge potential for the production of value added products.

Glucoamylase is an important enzyme for starch hydrolysis due to its catalytic effect of releasing glucose from the non-reducing ends of starch [14]. It is widely used in the food, feed and pharmaceutical industries, mainly for the production of glucose syrup, high fructose corn syrup, and alcohol. Traditionally, filamentous fungi were used for producing glucoamylase, with *Aspergillus niger*, *Aspergillus awamori* and *Rhizopus oryzae* being the major strains used for commercial gluco-amylase production [15].

To reduce the production costs, various agriculture residues have been explored for gluco-amylase production, including wheat bran, green gram bran, black gram bran, corn flour, barley flour, maize bran, rice bran, rice flakes and food waste. Table 1 lists some recent studies on gluco-amylase production together with the fermentation conditions used in these reports.

Table 1. Recent studies on glucoamylase production via fungal fermentation.

Substrate	Strain	Fermentation Type	Gluco-Amylase Production	Ref.
Babassu cake (kernel residue)	*A. awamori*	SSF/4 days	22.8 U/mL	[16]
Babassu cake, castor seed, sunflower & canola cakes	*A. awamori, A. wenti, P. verrucosum*	SSF/4 days	29.8 U/g	[17]
Kitchen waste/Wheat bran	*A. niger*	SSF/5 days	1838 U/g	[18]
Pastry waste & mixed food waste	*A. awamori*	SSF/10 days	76.1 ± 6.1 U/mL	[19]
Rice bran	*A. awamori, A. niger, A. terreus, A. tamarii*	SmF	264.5 U/g	[20]
Sorghum pomace	*A. niger* and *Saccharomyces cerevisae*	SmF/3 days	3.3 U/mg protein	[21]
Waste bread	*A. awamori*	SSF/4–5 days	114–130.8 U/g	[22, 23]
Waste potato mash	*A. niger*	SSF	274.4 U/mL	[24]
Wheat bran	*A. awamori*	SSF/4 days	9157 U/g	[25]
Wheat bran	*A. niger*	SSF/4 days	1.345 ± 0.009 IU/mL/min	[26]
Wheat milling by-product	*A. awamori*	SSF/4 days	48 U/g 4.4 U/mL	[27]

SSF: Solid state fermentation; SmF: Submerged fermentation; U/g: Unit per gram dry weight biomass.

Media composition and growth conditions were reported to influence glucoamylase production significantly. At low concentrations, glucose has been reported to be an inhibitor for the production of glucoamylase, while some nitrogen sources such as yeast extract, ammonium sulphate, ammonium nitrate, urea, meat extract and peptone have been used to promote glucoamylase production [28,29]. Optimization of fermentation either using one factor at a time design [20,22,23] or using response surface method [24] is still a main approach to improve glucoamylase production, along with gluco-amylase producing strain selection and genetic modification [20,24]. It has been shown that a 24% increase in glucoamylase activity was achieved through optimization of SSF media and parameters by *A. oryzae* using agro residues as substrate [30].

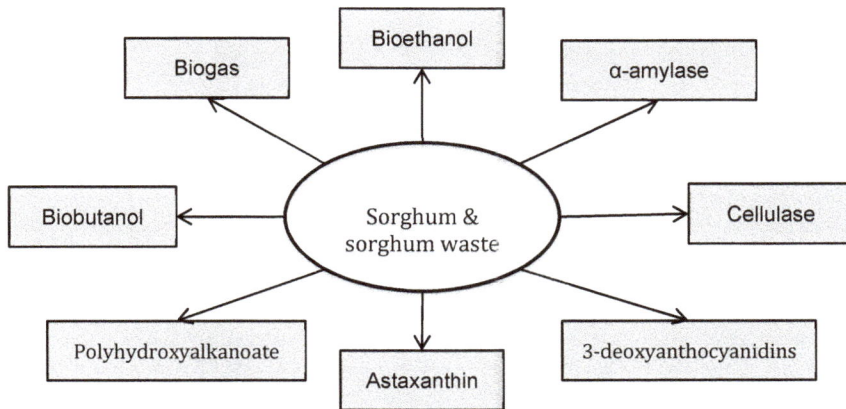

Figure 1. The schematic diagram showing the potential food ingredient and non-food products that can be derived from sorghum and sorghum waste.

Although sorghum bran is one of the major agriculture waste streams, and contains high residual starch (up to 53% (w/w) [7]), the utilization of sorghum bran has not been fully investigated. In this study, a sorghum bran-based biorefining concept has been developed for glucoamylase production and using the resulting glucoamylase to hydrolyze sorghum bran to produce a sugar-rich generic fermentation medium. Glucoamylase synthesis using *Aspergillus awamori* via Submerged Fermentation (SmF) was carried out. Fermentation conditions, such as substrate concentration, pH and temperature were investigated using one factor at a time design. Bench top fermentation for relatively large-scale enzyme production was carried out to produce sufficient amount of crude enzyme, which was subsequently used for the hydrolysis of sorghum bran and compared with commercial enzymes. Finally, the economic benefit of utilizing sorghum bran for the production of glucoamylase was discussed briefly.

2. Materials and Methods

2.1. Sorghum Bran

The sorghum (*Sorghum bicolor*) used in this study is a variety of red sorghum, which was purchased from a local market in Ikorodu, Lagos State, Nigeria in Spring 2017. The sorghum was subjected to three different milling processes using smart peanut butter maker (wet milling), blender (wet milling) and knife mill (dry milling). For wet milling, sorghum was steeped in tap water (2:5 w/v) for 3 days at room temperature and was wet milled using either a smart peanut butter maker (Smart@, Nostalgia, Amazon, UK) or a food-processing blender (Cookworks, Argos, UK). The milled biomass was sieved with muslin cloth to remove the starch component from the slurry. The remaining component sorghum bran (consisting of the outer layers of the cereal grain and residual starch) was dried in an oven at 60 °C for 3 days. For milling using a knife mill, air dried sorghum grain was milled by a lab knife mill to pass through a 2-mm sieve screen. The resulting product was then subsequently sieved using a 1-mm sieve. The particles above 1-mm were used as sorghum bran. The total starch content of the sorghum bran sample was determined using an enzymatic starch analysis kit (Megazyme®, Bray, Ireland).

2.2. Microorganisms

The strain used for glucoamylase production was *Aspergillus awamori* 2B.361 U2/1 (*A. awamori*), which was kindly provided by Prof. Colin Webb, the University of Manchester (UK). Procedures for storing, and cultivating *A. awamori* were described by Koutinas et al. [31].

2.3. Sorghum Bran Submerged Fermentation

Submerged fermentation was carried out both in shake flask for preliminary assessment and in 2-L fermenters (Electrolab FerMac 360, UK) for large-scale enzyme production. The sorghum bran concentration used in the initial fermentation was 4% (w/v). 250 mL shaking flasks were used in all experiments except the study on the impact of medium loading ratio. The working volume was 100 mL unless specified. Several drops of silicon antifoam (0.002% v/v) were added to the complex medium in order to prevent foaming. Unless specified, no other nutrient or chemical were added into the fermentation media. The media were sterilised at 121 °C for 15 min and allowed to cool down before *A. awamori* was added at an inoculation ratio of 1×10^7 spores/g dry weight sorghum bran. The mixture was fermented in a shaking incubator (Incu-Shake FL24-1R, SciQuip, UK) at 28 °C and 200 rpm. Glucoamylase production was investigated using different conditions, including initial pH (3.0–8.0), temperature (26 °C, 28 °C and 30 °C), substrate concentration (2–10% w/v), medium loading ratio (50–250 mL in 500 mL bottle), yeast extract (0–10 g/L, Sigma-Aldrich, UK), and with and without the addition of minerals. Five mineral solutions were explored in this study, as shown in Table 2. All SmF were carried out in triplicates.

Table 2. Mineral composition investigated in SmF.

Mineral Solution	Composition	REF
A	$(NH_4)_2SO_4$ 1 g/L, KH_2PO_4 0.5 g/L, K_2HPO_4 0.5 g/L, $MgSO_4$ 0.2 g/L	[32]
B	KH_2PO_4 6 g/L, $MgSO_4 \cdot 7H_2O$ 1 g/L, $FeCl_3 \cdot 4H_2O$ 10 mg/L	[33]
C	KH_2PO_4 1 g/L, $MgSO_4 \cdot 7H_2O$ 0.3 g/L, $CaCl_2$ 0.3 g/L	[34]
D	KH_2PO_4 1 g/L, $MgSO_4 \cdot 7H_2O$ 0.5 g/L	Designed in this study
E	NH_4NO_3 1.5 g/L, K_2HPO_4 2.5 g/L, KH_2PO_4 1.5 g/L, $MgSO_4$ 0.12 g/L, NaCl 0.25 g/L	Designed in this study

For large-scale glucoamylase production, strain *A. awamori* was cultured in 250 mL shaking flasks containing 50 mL inoculation medium. The inoculation medium contained 40 g/L of glucose and 10 g/L of yeast extract. The seed fermentation was carried out at 28 °C, 200 rpm for 3 days in the shaking incubator (SciQuip Incu-Shake FL24-1R). The fermentation medium contained 10% (w/v) sorghum bran, 2.5 g/L yeast extract and was prepared with mineral solution C. 1 mL of sterilized silicon antifoam (0.02%, v/v) was added to the fermentation medium before inoculation. The fermentation was carried out at 28 °C, 500 rpm and an aeration rate at 1.0 L/min. The pH was controlled to 6.0 by adding 2 M NaOH solution or 2 M HCl solution. 10 mL of sample were taken at different time intervals for glucoamylase analysis.

2.4. Glucoamylase Enzyme

Glucoamylase activity was measured using the method described by Bernfeld [35]. Two mL of fermentation sample was spun in a centrifuge (Eppendorf, UK) at 5000 rpm for 5 min to remove cell. The liquid suspension was then used as crude enzyme solution. A reaction mixture containing 0.9 mL of 0.05 mM citrate buffer (pH 5.0), 1.0 mL autoclaved starch solution (1%, w/v) and 0.1 mL of crude enzyme solution was incubated at 50 °C for 20 min. Then 3 mL of 3,5-dinitrosalicyclic acid (DNSA) reagent [36] was added to the incubated mixture. The reaction mixture was heated in a vigorously boiling water bath for 5 min and was allowed to cool. Absorbance was measured using a spectrophotometer at 540 nm. Glucoamylase activity unit (U) was express as the amount of enzyme releasing one μmole of glucose equivalent per minute under the assay condition and enzyme activity was express in terms of units per mL (U/mL).

2.5. Enzymatic Hydrolysis of Sorghum Bran

The enzymatic hydrolysis of sorghum bran was initiated by gelatinizing a mixture of 4 g sorghum bran in 50 mL of deionised water in a boiling water bath at 100 °C for 20 min in a 250 mL conical flask. The agitation was carried out using a glass rod to mix for 30 s every 5 min. After gelatinisation, the substrate was cooled to 55 °C. Then, either crude gluco-amylase enzyme solution or commercial enzymes (gluco-amylase and α-amylase from Megazyme@) were added at an enzyme loading ratio of 50 U/g. The hydrolysis was carried out in a shaking incubator (SciQuip Incu-Shake MIDI), 200 rpm at 55 °C for 48 h. The sorghum bran hydrolysis yield was calculated using the following equation:

$$Hydrolysis\ yield = \frac{Weight\ of\ glucose}{Weight\ of\ bran \times starch\ content \times 1.11} \qquad (1)$$

2.6. Sugar Analysis

The amounts of sugars were quantified by HPAEC-PAD. The sample was filtered through a 0.2 μm syringe filter and was then transferred into a 1.5 mL agilent auto sampler vial. The monosaccharides were analysed using a Dionex ICS-3000 Reagent-Free™ Ion Chromatography on a with Dionex ICS-3000 system, electrochemical detection using ED 1 and computer controller. A CarboPacTM PA 20 column (3 × 150 mm/; Dionex, Sunnyvale, CA, USA) was used and the mobile phase was 10 mM NaOH with a flow rate of 0.3 mL/min. The injection volume was 25 μL and the column temperature was 30 °C.

2.7. Statistical Analysis

Microsoft Excel (2013) was used to calculate the results obtained from all the experiment such as the standard deviation.

3. Results and Discussion

3.1. Starch Content in the Sorghum Brans

Three milling processes using a peanut butter maker (wet milling), blender (wet milling) and knife mill were examined for separating starch from sorghum bran. The starch contents of the resulting sorghum bran were analyzed as shown in Table 3.

Table 3. The starch content in sorghum bran.

Sorghum species	Milling Processing	Starch Content (w/w)	REF
Red sorghum	Peanut butter maker	16.4 ± 1.3%	This study
Red sorghum	Blender	13.0 ± 0.8%	This study
Red sorghum	Knife mill	81.9 ± 3.2%	This study
Red sorghum	A tangential abrasive dehulling device	30%	[37]
Red sorghum	Buhler mill/hammer mill	24%	[4]
Red sorghum	Wet milling	52.96 ± 1.43%	[7]
White sorghum	Wet milling	49.7 ± 0.86%	[7]

This revealed that extremely high starch (81.93%, w/w) was retained in the bran obtained after knife mill (dry milling), indicating that dry milling of sorghum grain in the lab was not suitable for separating the bran from the kernel. By contrast, wet milling using a peanut butter maker and the blender mill successfully isolated bran from sorghum kernel, resulting in sorghum brans containing only 16.4% and 13.0% starch, respectively. In comparison with other reports of starch composition in sorghum bran (Table 3), the starch concentrations obtained in this study were much lower, suggesting wet milling using a peanut butter maker or a blender was a suitable technology for sorghum starch recovery.

3.2. Glucoamylase Production Using Sorghum Bran Via SmF

The wet milled sorghum bran derived using the peanut butter maker was used for the production of glucoamylase in a submerged fermentation (SmF). The initial fermentation was carried out using 4% (w/v) sorghum bran and no addition of nutrients. As shown in Figure 2A, an increase in glucoamylase activity was detected until 120 h, then the enzyme activities decreased sharply. The peak glucoamylase activity was 1.90 U/mL.

To improve glucoamylase production, the sorghum bran was augmented with a mineral solution, containing K_2HPO_4 2.5 g/L, NH_4NO_3 1.5 g/L, KH_2PO_4 1.5 g/L, $MgSO_4$ 0.12 g/L and NaCl 0.25 g/L (Mineral solution E in Table 2), respectively. Adding mineral to the sorghum bran-containing fermentation medium increased the glucoamylase activity from 1.90 U/mL to 3.60 U/mL and reduced the time requirement for the peak enzymatic activity from 120 h to 72 h (Figure 2A). Further to this result, the supplement of four more mineral solutions were explored (Table 2), and the results are shown in Figure 2B. The mineral solution C led to the highest glucoamylase activity of 5.03 U/mL, which was then used in the following fermentation experiments. Comparing mineral solution C and mineral solution D, the removal of $CaCl_2$ led to a significant decrease of glucoamylase activity, indicating that the calcium may play a key role in glucoamylase production. However, further experiment is required to confirm this hypothesis.

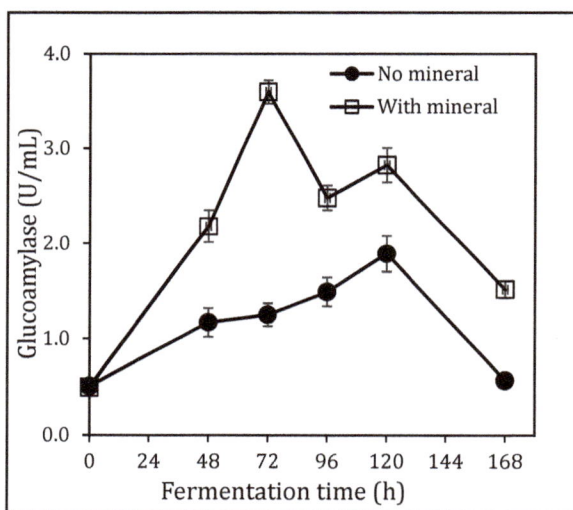

(A)

Figure 2. *Cont.*

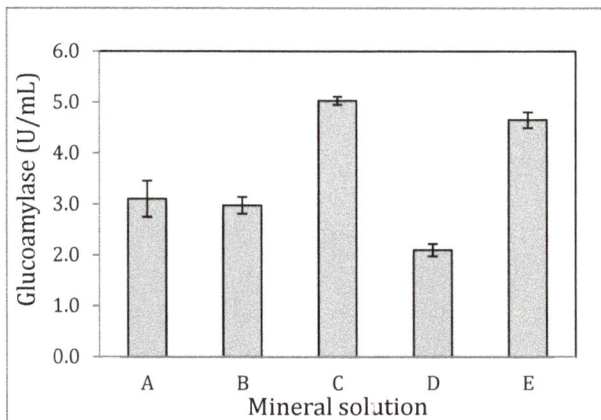

(B)

Figure 2. The effect of mineral solution on glucoamylase production in SmF. (**A**): the comparison of gluco-amylase production profile in SmF with and without mineral solution E. (**B**): Gluco-amylase production using mineral solutions A–E, fermentation time 72 h, temperature 28 °C, 4% (w/v) sorghum bran, no extra nutrients, no pH adjustment.

3.3. Impact of Substrate Concentration and pH on Glucoamylase Production

The substrate concentration determines the availability of carbon source in the fermentation system. The low glucoamylase activities observed in the above experiment could be due to an insufficient supply of carbon sources. The effect of sorghum bran concentration on glucoamylase activities in SmF was explored (2%, 4%, 6%, 8%, and 10%, w/v). The utilisation of 6% starch-rich sorghum milling product (SS) from peanut maker milling process was also included as a comparison. The starch-rich sorghum milling product contained 49.4% (w/w) starch, corresponding to ~18% sorghum bran concentration based on the total starch content. As shown in Figure 3A, glucoamylase activity increased as the substrate concentration increased. The highest glucoamylase activity of 12.6 ± 0.2 U/mL was obtained when 10% sorghum bran was used. The starch-rich sorghum milling product at 6% resulted in a similar glucoamylase activity (6.4 ± 1.8 U/mL) as that observed with 8% sorghum bran (6.2 ± 1.1 U/mL). These results indicated that sorghum bran contained some nutrients other than starch to facilitate glucoamylase synthesis. Further increasing substrate concentration was not carried out mainly due to the difficulty of mixing problem at high sorghum bran concentration in shake flasks.

The effect of initial pH on glucoamylase production was investigated in SmF as shown in Figure 3B. A gradual increase in enzyme production was observed at pH 3.0, 4.0, 5.0 and 6.0, peak enzyme production obtained was 8.7 ± 0.8 U/mL, 16.9 ± 0.3 U/mL, 16.5 ± 0.4 U/mL, and 19.3 ± 0.5 U/mL, respectively. At pH 7, the peak enzyme activity was obtained at 48 h (11.9 ± 0.4 U/mL) before a declining trend was observed. There was a slow increase in glucoamylase activity at pH 8.0 for 96 h before a peak enzyme activity was observed at 120 h (8.5 ± 0.5 U/mL).

The optimum pH for glucoamylase enzyme production was determined to be pH 6.0 after 72 h of the fermentation. These results agreed with previous results that 3 days of cultivation led to a better glucoamylase accumulation. Although a higher amylase synthesis was reported at pH 8.0 by *Bacillus sp* under SSF [38], the majority of research has reported that the best pH for amylase production was approximately pH 6.0 [39,40].

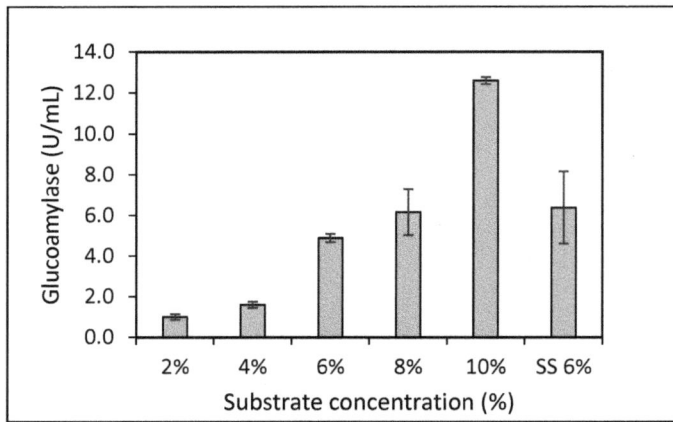

(A)

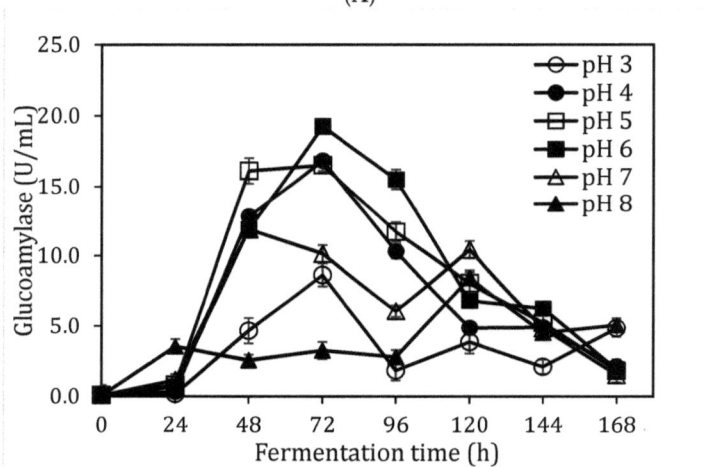

(B)

Figure 3. The effect of substrate concentration (**A**) and pH (**B**) on glucoamylase production. For the fermentation in (**3A**): fermentation time 72 h, temperature 28 °C, with mineral solution C, no extra nutrients, no pH adjustment. For the fermentation in 3B: temperature 28 °C, 10% (w/v) sorghum bran, with mineral solution C, no extra nutrients, no pH adjustment.

3.4. Impact of Medium Loading Ratio on Glucoamylase Production

The fermentation medium loading ratio affected the dissolved oxygen level and the mixing pattern in the shake flask. The dissolved oxygen level had an important impact on the physiology and metabolism of the microorganism. At a low level of oxygen supply, the production of essential enzymes was inhibited, while at a high aeration rate could have detrimental effects on the growth of microorganism and subsequent enzyme production. Mechanical mixing in a bioreactor determined the heat and mass transfer rates, therefore, impacting cell growth and enzyme synthesis. The impact of medium loading ratio on glucoamylase production was determined in SmF using 500 mL shake flasks at 200 rpm, 28 °C.

As shown in Figure 4, there was no significant increase in glucoamylase production when the aeration ratios of 50/500 mL or 100/500 mL were used. There was an increase in glucoamylase activity and a peak increase was obtained after 72 h of fermentation when 150/500 mL and 200/500 mL (11.1 ± 0.2 U/mL and 12.7 ± 0.3 U/mL) was used, respectively, while a peak increase in glucoamylase was obtained at 96 h when medium loading ratio of 250/500 mL (11.9 ± 0.3 U/mL) was used. It was expected that a low medium loading ratio would encourage oxygen transfer and thus benefit gluco-amylase production. However, the results in Figure 4 clear indicated a higher loading ratio at 200/500 mL was a preferred condition. This could attribute to the high viscosity in sorghum bran derived medium, which created significant mixing difficulty when the actual reaction volume was low. The insufficient mixing subsequently affected cell growth and glucoamylase production.

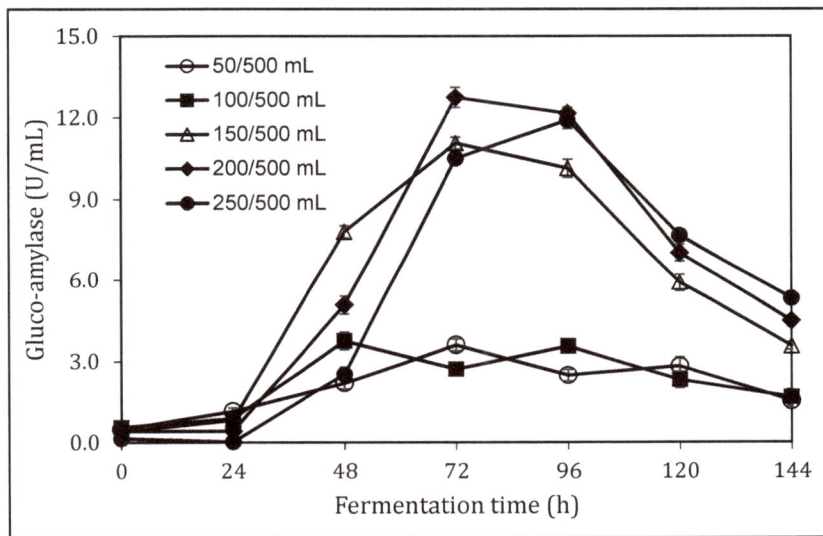

Figure 4. The effect of medium loading ratio on glucoamylase production. Fermentation was carried out at 28 °C, 10% (w/v) sorghum bran, with mineral solution C, at pH 6.0.

3.5. Impact of Yeast Extraction and Temperature on Glucoamylase Production

Yeast extract is used as a nitrogen and nutrient source in many bacterial culture media. Yeast extract contains an abundance of vitamins, minerals and amino acids, which are necessary for cell growth and enzyme synthesis. The addition of yeast extract was carried out with the aim to further increase glucoamylase enzyme production (Figure 5A). The optimum glucoamylase activity (13.0 ± 0.3 U/mL) was obtained after 3 days of fermentation with 2.5 g/L yeast extract addition.

Temperature has an important effect on enzyme production, as a reaction rate generally increases with temperature to a maximal level before a decline occurs with further increase in temperature due to the enzymes' susceptibility to denaturation. The SmF was explored using temperatures ranging from 26 °C to 30 °C. As shown in Figure 5B, fermentation at 28 °C had the most significant glucoamylase activity after 96 h of fermentation (10.8 ± 0.5 U/mL). At 30 °C, the gluco-amylase activity showed a slow gradual increase up to 120 h, while at 26 °C the glucoamylase activity had a similar pattern as observed at 28 °C, but with less glucoamylase accumulation. This results obtained in this study agreed with Khan and Yadav [39], which reported an optimum α-amylase production at 28 °C for *A. niger*. Maximum amylase production by *A. niger* and *R. stolonifera* was achieved at 30 °C [41,42].

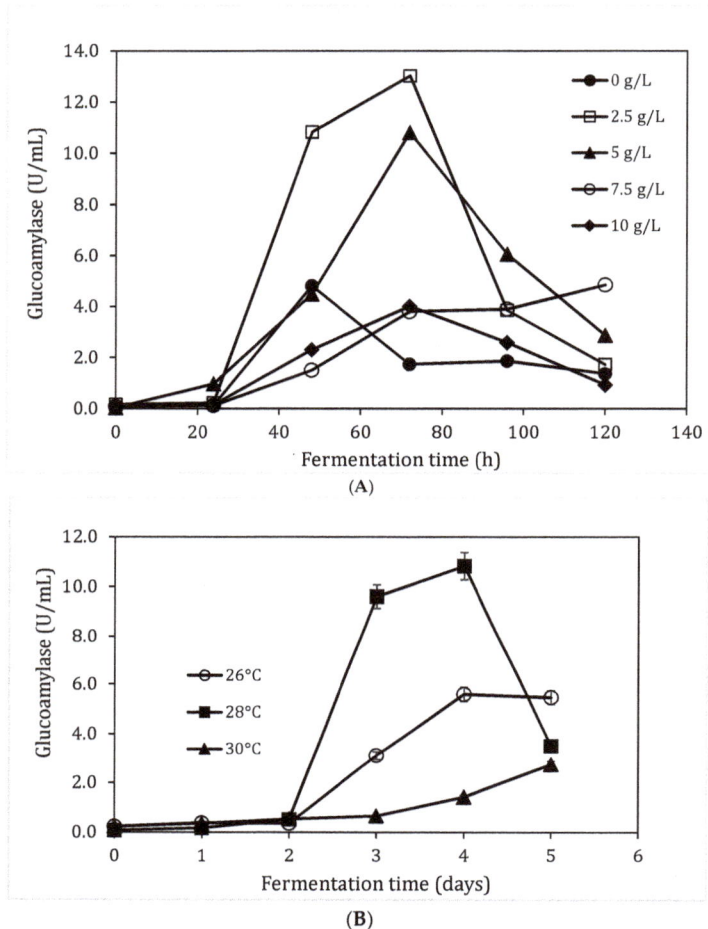

Figure 5. The effect of yeast extract (**A**) and temperature (**B**) on glucoamylase production in SmF. Fermentation was using 10% (w/v) sorghum bran, with mineral solution C at pH 6.0.

3.6. Glucoamylase Production in Bench Top Fermenters

The above results indicated that the highest glucoamylase activity was achieved using a 10% (w/v) sorghum bran loading ratio, with addition of mineral solution C, 2.5 g/L yeast extract, at pH 6.0, 28 °C, and a liquid loading ratio of 200 mL in 500 mL shaking flasks. The fermentation conditions were used for the SmF of sorghum bran in 2-L fermenters. The scale up was repeated in four batches, and glucoamylase activities were in the range of 20.7 U/mL to 23.5 U/mL. A typical glucoamylase production profile was shown in Figure 6. Since bench top fermenters provide sufficient mixing at high substrate loading ratios, fermentations with high sorghum bran loading ratios of 12.5% and 15% (w/v) were also investigated. Typical glucoamylase production profiles are shown in Figure 6. When the substrate concentration was increased to 12.5%, glucoamylase production was enhanced to 37.6 U/mL (corresponding to 250 U/g dry weight sorghum bran), but the fermentation time was extended to 115 h. A sharp decline in enzyme activity at 120 h was observed due to foaming in the fermenter as a result of fungal autolysis. High stirring speed at 500 rpm was used due to high viscosity of the fermentation medium at high sorghum bran loading ratio. Vigorous agitation increased oxygen transfer and nutrient transfer, but resulted in mechanical stress, excessive foaming, disruption and physiological disturbance of cells. Further increasing substrate concentration to 15% did not lead to an improved glucoamylase production, mainly due to the insufficient mass transfer in the fermenter.

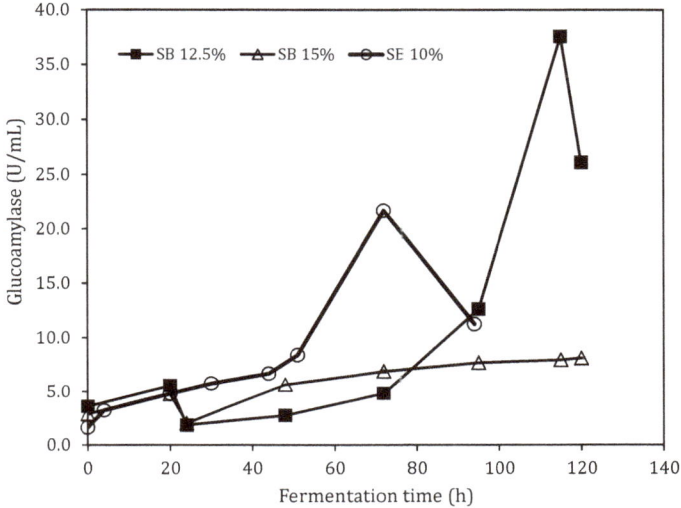

Figure 6. Typical glucoamylase accumulation profiles in 2-L fermenters using 10%, 12.5% and 15% (w/v) sorghum bran. Fermentation was carried out at 28 °C, 500 rpm, air aeration rate at 1.0 L/min, with mineral solution C at pH 6.0.

The 2-L fermentation results confirmed that the optimum fermentation condition conditions obtained in shake flask experiments were valid. In comparison with the glucoamylase activities reported in the literature (Table 1), the glucoamylase activity of 250 U/g was close to that obtained using rice bran (264.5 U/g) [20], and was among the high glucoamylase activities reported. The results indicated that sorghum bran was a suitable substrate for the production of glucoamylase. Further more, the high enzyme concentration in the crude enzyme solution facilitates the following enzymatic hydrolysis, as it allows a higher concentration of substrate to used in the hydrolysis step.

3.7. Sorghum Bran Hydrolysis Using Crude Glucoamylase Solution

The utilisation of the crude glucoamylase for the hydrolysis of sorghum bran was carried out and compared with commercial enzymes (glucoamylase and α-amylase, from the Megazyme@ starch kit). The substrate loading ratio was 80 g/L, the enzyme loading ratio was 50 U/g dry weight sorghum bran and the hydrolysis was carried out at 55 °C for 120 h. By the end of the hydrolysis, the glucose concentration in the hydrolysis of the crude enzyme and the commercial enzyme were 11.32 ± 0.8 g/L and 11.74 ± 0.5 g/L, respectively, corresponding to a hydrolysis yield of 78.7% and 81.6% of the theoretical yield, respectively. In order to improve sugar content in the hydrolysate, a solid loading ratio of 200 g/L was carried out using a 2 L fermenter, at 55 °C, 500 rpm for 48 h. Around 700 mL of sorghum bran hydrolysate was obtained, with a glucose concentration of 38.7 ± 1.3 g/L, corresponding to 94.1% of the theoretical hydrolysis yield.

3.8. Economic Evaluation of Glucoamylase Production in a Bioethanol Production Process

Ahmed El-Iman et al. recently developed a model for estimating bioethanol production potential in Nigeria [7]. In the model, an acid hydrolysis process was used, and the starch content in the sorghum bran was determined to be 52.96% (w/w). Some 0.73 million tons of bioethanol was estimated to be produced from 7.56 million tons of available sorghum crop [7]. If the acid hydrolysis process could be replaced by the integrated biorefining strategy reported in this study, a 10% increase in bioethanol production could be achieved (Figure 7A). As shown in Figure 7A, although 17% of the sorghum bran was used for glucoamylase production, the significant increase in hydrolysis yield (94.1%) led to 0.81 million tons of bioethanol being produced. This suggested that using sorghum bran for on-site glucoamylase hydrolysis would economically benefit the bioethanol production process. Inthe case that the sorghum bran contains 16.4% starch (Table 3), only 0.25 million tons of bioethanol could be produced (Figure 7B). In this scenario, it would be more economically feasible to use all available sorghum bran for glucoamylase production as enzymes are relatively higher value products than bioethanol, and the capital investigation for glucoamylase production would be lower than that for bioethanol production.

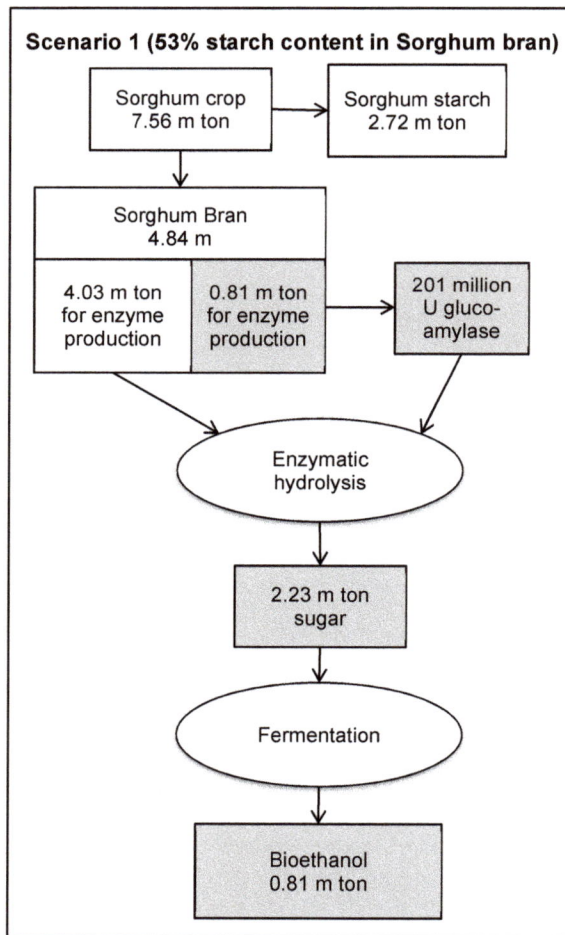

(A)

Figure 7. *Cont.*

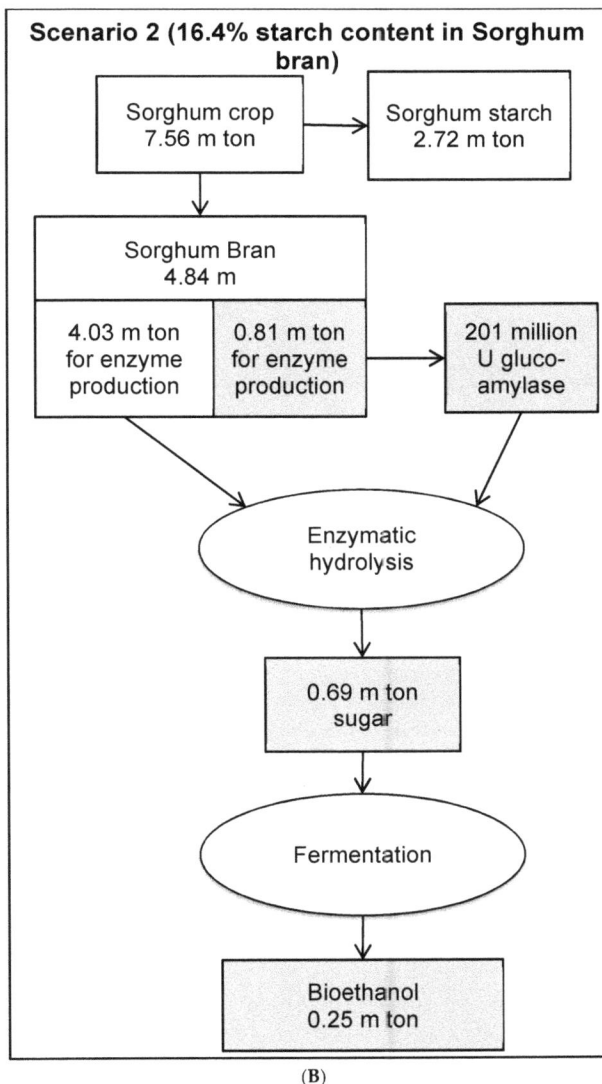

Figure 7. Mass balance of a sorghum bran based biorefining strategy for bioethanol production via glucoamylase hydrolysis process. (**A**): scenario 1, the starch content in sorghum bran was 53% [7]; (**B**) scenario 2, the starch content in sorghum bran was 16.4% (this study).

4. Conclusions

In this study, the utilization of sorghum bran for the glucoamylase production was investigated. The compositional analysis indicated that sorghum bran derived from wet milling using a peanut butter maker contained 16.4% (w/w) starch. Investigation of fermentation conditions led to a 10-fold increase in glucoamylase production from 1.90 U/mL to 19.3 U/mL. Fermentations using 2-L fermenters confirmed the shake flask experimental results. Further increase in the substrate concentration to 12.5% (w/v) in 2-L fermentations achieved a glucoamylase concentration of 37.6 U/mL. The utilization of the

Foods **2019**, *8*, 279

enzyme solution for the hydrolysis of sorghum bran indicated that the crude enzyme was comparable with commercial enzymes. A sorghum bran hydrolysate containing 38.7 g/L glucose was obtained, which can be used as a generic fermentation feedstock for the fermentative production of biofuels and biochemicals.

Author Contributions: Investigation, O.M.; Supervision, D.G. and C.D.; Writing – original draft, O.M.; Writing – review & editing, O.M., D.G., X.Z. and C.D.

Funding: This research received no external funding.

Acknowledgments: The authors acknowledge the fund from the University of Huddersfield, under the programme of URF (URF2015/24).

Conflicts of Interest: The authors declare no conflict of interest.

References

1. Fuller, M.F. Chapter title? In *Encyclopedia of Farm Animal Nutrition*; CABI: Wallingford, UK, 2014.
2. Beta, T.; Chisi, M.; Monyo, E.S. Sorghum/harvest, storage and transport. In *Encyclopedia of Grain Science*; Wingley, C.W., Corke, H., Walker, C.E., Eds.; Elsevier: Oxford, UK, 2004; pp. 119–126.
3. Waniska, R.D.; Rooney, L.W.; McDonough, M. Sorghum/utilization. In *Encyclopedia of Grain Science*; Wingley, C.W., Corke, H., Walker, C.E., Eds.; Elsevier: Oxford, UK, 2004; pp. 126–138.
4. Adeyemi, I.A. Dry-milling of sorghum for ogi manufacture. *J. Cereal Sci.* **1983**, *1*, 221–227. [CrossRef]
5. Li, S.Z.; Li, G.M.; Zhang, L.; Zhou, Z.; Han, B.; Hou, W.; Wang, J.; Li, T. A demonstration study of ethanol production from sweet sorghum stems with advanced solid state fermentation technology. *Appl. Energy* **2013**, *102*, 260–265. [CrossRef]
6. Jiang, D.; Hao, M.; Fu, J.; Liu, K.; Yan, X. Potential bioethanol production from sweet sorghum on marginal land in China. *J. Clean Prod.* **2019**, *220*, 225–234. [CrossRef]
7. Ahmed El-Imam, A.; Greetham, D.; Du, C.; Dyer, P. The development of a biorefining strategy for the production of biofuel from sorghum milling waste. *Biochem. Eng. J.* **2019**, *150*, 107288. [CrossRef]
8. Cai, D.; Dong, Z.; Han, J.; Yu, H.; Wang, Y.; Qin, P.; Wang, Z.; Tan, T. Co-generation of bio-butanol and bio-lipids under a hybrid process. *Green Chem.* **2016**, *18*, 1377–1386. [CrossRef]
9. Nozari, B.; Mirmohamadsadeghi, S.; Karimi, K. Bioenergy production from sweet sorghum stalks via a biorefinery perspective. *Appl. Microbiol. Biotechnol.* **2018**, *102*, 3425–3438. [CrossRef] [PubMed]
10. Islam, M.S.; Guo, C.; Liu, C.-Z. Enhanced hydrogen and volatile fatty acid production from sweet sorghum stalks by two-steps dark fermentation with dilute acid treatment in between. *Int. J. Hydrogen Energy* **2018**, *43*, 659–666. [CrossRef]
11. Stoklosa, R.J.; Johnston, D.B.; Nghiem, N.P. Utilization of Sweet Sorghum Juice for the Production of Astaxanthin as a Biorefinery Co-Product by *Phaffia rhodozyma*. *ACS Sustain. Chem. Eng.* **2018**, *6*, 3124–3134. [CrossRef]
12. Vanamala, J.K.P.; Massey, A.R.; Pinnamaneni, S.R.; Reddivari, L.; Reardon, K.F. Grain and sweet sorghum (*Sorghum bicolor* L. Moench) serves as a novel source of bioactive compounds for human health. *Crit. Rev. Food Sci. Nutr.* **2018**, *58*, 2867–2881. [CrossRef]
13. Lolasi, F.; Amiri, H.; Asadollahi, M.A.; Karimi, K. Using sweet sorghum bagasse for production of amylases required for its grain hydrolysis via a biorefinery platform. *Ind. Crop Prod.* **2018**, *125*, 473–481. [CrossRef]
14. Kumar, P.; Satyanarayana, T. Microbial glucoamylases: Characteristics and applications. *Crit. Rev. Biotechnol.* **2009**, *29*, 225–255. [CrossRef] [PubMed]
15. Norouzian, D.; Akbarzadeh, A.; Scharer, J.M.; Moo Young, M. Fungal glucoamylases. *Biotechnol. Adv.* **2006**, *24*, 80–85. [CrossRef] [PubMed]
16. López, J.A.; Lázaro, C.D.C.; Castilho, L.D.R.; Freire, D.M.G.; Castro, A.M.D. Characterization of multienzyme solutions produced by solid-state fermentation of babassu cake, for use in cold hydrolysis of raw biomass. *Biochem. Eng. J.* **2013**, *77*, 231–239. [CrossRef]
17. Castro, A.M.D.; Carvalho, D.F.; Freire, D.M.G.; Castilho, L.D.R. Economic analysis of the production of amylases and other hydrolases by Aspergillus awamori in solid-state fermentation of babassu cake. *Enzym. Res.* **2010**, *2010*, 576872. [CrossRef] [PubMed]

18. Wang, X.Q.; Wang, Q.H.; Liu, Y.Y.; Ma, H.Z. On-site production of crude glucoamylase for kitchen waste hydrolysis. *Waste Manag. Res.* **2009**, *28*, 539–544. [CrossRef] [PubMed]

19. Lam, W.C.; Pleissner, D.; Lin, C.S.K. Production of fungal glucoamylase for glucose production from food waste. *Biomolecules* **2013**, *3*, 651–661. [CrossRef] [PubMed]

20. Abdalwahab, S.A.; Ibrahim, S.A.; Dawood, E.S. Culture condition forthe production of glucoamylase enzyme by different isolates of Aspergillus spp. *Int. Food Res. J.* **2012**, *19*, 1261–1266.

21. Abu, E.A.; Ado, S.A.; James, D.B. Raw starch degrading amylase production by mixed culture of Aspergillus niger and Saccharomyces cerevisae grown on sorghum pomace. *Afr. J. Biotechnol.* **2005**, *4*, 785–790.

22. Melikoglu, M.; Lin, C.S.K.; Webb, C. Stepwise optimisation of enzyme production in solid state fermentation of waste bread pieces. *Food Bioprod. Process.* **2013**, *91*, 638–646. [CrossRef]

23. Melikoglu, M.; Lin, C.S.K.; Webb, C. Solid state fermentation of waste bread pieces by Aspergillus awamori Analysing the effects of airflow rate on enzyme production in packed bed bioreactors. *Food Bioprod. Process.* **2015**, *95*, 63–75. [CrossRef]

24. Izmirliogiu, G.; Demirci, A. Strain selection and media optimization for glucoamylase production from industrial potato waste by Aspergillus niger. *J. Sci. Food Agric.* **2016**, *96*, 2788–2795. [CrossRef] [PubMed]

25. Negi, S.; Banerjee, R. Optimization of extraction and purification of gluco-amylase produced by *A. awamori* in SSF. *Biotechnol. Bioprocess. Eng.* **2009**, *14*, 60–66. [CrossRef]

26. Imran, M.; Asad, M.J.; Gulfraz, M.; Qureshi, R.; Gul, H.; Manzoor, N.; Choudhary, A.N. Glucoamylase production from Aspergillus niger by using solid state fermentation process. *Pak. J. Bot.* **2010**, *44*, 2103–2110.

27. Du, C.; Lin, S.K.C.; Koutinas, A.; Wang, R.; Dorado, P.; Webb, C. A wheat biorefining strategy based on solid-state fermentation for fermentative production of succinic acid. *Bioresour. Technol.* **2018**, *99*, 8310–8315. [CrossRef] [PubMed]

28. Pavezzi, F.C.; Gomes, E.; Silva, R.D. Production and characterization of glucoamylase from fungus *Aspergillus awamori* expressed in yeast Saccharomyces cerevisiae using different carbon sources. *Braz. J. Microbiol.* **2008**, *39*, 108–114. [CrossRef] [PubMed]

29. Kumar, P.; Satyanarayana, T. Optimization of culture variables for improving glucoamylase production by alginate-entrapped Thermomucor indicae-seudaticae using statistical methods. *Bioresour. Technol.* **2007**, *98*, 1252–1259. [CrossRef] [PubMed]

30. Zambare, V. Solid state fermentation of Aspergillus oryzae for gluco-amylase production on agro residues. *Int. J. Life Sci.* **2010**, *4*, 16–25. [CrossRef]

31. Koutinas, A.; Belafi-Bako, K.; Kabiri-Badr, A.; Toth, A.; Gubicza, L.; Webb, C. Enzymatic hydrolysis of polysaccharides, Hydrolysis of starch by an enzyme complex from fermentation by Aspergillus awamori. *Food Bioprod. Process.* **2001**, *79*, 41–45. [CrossRef]

32. Pensupa, N.; Jin, M.; Kokolski, M.; Archer, D.B.; Du, C. A solid state fungal fermentation-based strategy for the hydrolysis of wheat straw. *Bioresour. Technol.* **2013**, *149*, 261–267. [CrossRef]

33. Bancerz, R.; Osińska-Jaroszuk, M.; Jaszek, M.; Janusz, G.; Stefaniuk, D.; Sulej, J.; Jarosz-Wiłkołazka, A.; Rogalski, J. New alkaline lipase from Rhizomucor variabilis: Biochemical properties and stability in the presence of microbial EPS. *Biotechnol. Appl. Biochem.* **2016**, *63*, 67–76. [CrossRef]

34. Yang, S.Q.; Xiong, H.; Yang, H.Y.; Yan, Q.J.; Jiang, Z.Q. High-level production of beta-1, 3–1, 4-glucanase by Rhizomucor miehei under solid-state fermentation and its potential application in the brewing industry. *J. Appl. Microbiol.* **2015**, *118*, 84–91. [CrossRef] [PubMed]

35. Bernfeld, P. Amylases alpha and beta. *Meth. Enzymol.* **1955**, *1*, 149–158.

36. Miller, G.L. Use of dinitrosalicylic acid reagent for determination of reducing sugar. *Anal. Chem.* **1959**, *31*, 426–428. [CrossRef]

37. Corredor, D.Y.; Bean, S.; Wang, D. Pretreatment and enzymatic hydrolysis of sorghum bran. *Cereal Chem.* **2007**, *84*, 61–66. [CrossRef]

38. Vijayaraghavan, P.; Kalaiyarasi, M.; Vincent, S.G.P. Cow dung is an ideal fermentation medium for amylase production in solid state fermentation. *J. Gene. Eng. Biotechnol.* **2015**, *13*, 111–117. [CrossRef]

39. Khan, J.A.; Yadav, S.K. Production of alpha amylase by Aspergillus niger using cheaper substrates employing solis state fermentation. *Int. J. Plant Anim. Environ. Sci.* **2011**, *13*, 100–108.

40. Saleem, A.; Ebrahim, M.K. Production of amylase by fungi isolated from legume seeds collected in Almadinah Almunawwarah, Saudi Arabia. *J. Taibah. Univ. Sci.* **2014**, *8*, 90–97. [CrossRef]

41. Simair, A.A.; Qureshi, A.S.; Khushk, I.; Ali, C.H.; Lashari, S.; Bhutto, M.A.; Mangrio, G.S.; Lu, C. Production and partial characterization of alpha amylase enzyme from Bacillus sp bcc 01–50 and potential application. *Biomed. Res. Int.* **2017**, *2017*, 9173040. [CrossRef]
42. Haqh, I.R.; Albdullah, A.; Shah, A.H. Isolation and screening of fungi for the biosynthesis of alpha amylase. *Biotechnology* **2002**, *12*, 61–66.

foods

MDPI

Article

Emulsifiers from Partially Composted Olive Waste

Aikaterini Koliastasi [1], Vasiliki Kompothekra [2], Charilaos Giotis [2,*], Antonis K. Moustakas [2], Efstathia P. Skotti [2], Argyrios Gerakis [2], Eleni Kalogianni [1] and Christos Ritzoulis [1]

[1] Department of Food Science and Technology, International Hellenic University, Sindos Campus, 57400 Thessaloniki, Greece
[2] Department of Food Science and Technology, Ionian University, Vergoti Avenue, 28100 Argostoli, Greece
* Correspondence: hgiotis@ionio.gr

Received: 12 June 2019; Accepted: 17 July 2019; Published: 20 July 2019

Abstract: Partial (one month) composting of solid olive processing waste is shown to produce extractable emulsifiers. Size exclusion chromatography (SEC) and Fourier-transform infra-red spectroscopy (FTIR) show that these consist of polysaccharides and proteins from the composted waste. Aqueous extraction at pH 5, pH 7, and pH 9 all yield extracts rich in oligosacchrides and oligopeptides which derive from the break-down of the macromolecules under composting, with the extract obtained at pH 5 being the richer in such components. Fourier-transform infra-red (FTIR) spectroscopy also confirms that these materials consist of proteinic and poly/oligosaccharidic populations. These materials can emulsify stable oil–in–water emulsions at pH 3 for a few days, while the same emulsions collapse in less than 24 h at pH 7. Confocal microscopy and droplet size distribution data suggest that Ostwald ripening, rather than coalescence, is the major course of emulsion instability. The above point to a short-process alternative to full composting in producing a high added value product from solid olive processing waste.

Keywords: emulsifier; olive waste; size exclusion chromatography (SEC); emulsion; compost; Ostwald ripening

1. Introduction

Olive oil is produced mainly in the Mediterranean area and EU countries with the average olive oil production in the EU in recent years being 2.2 million tons, representing around 80% of world production with half of it being produced in Spain [1]. Greece, Italy, and Spain account for about 97% of EU olive oil production [2]. Although olive oil has a positive effect on our health, its resulting by-products (olive mill waste) are acknowledged as a serious environmental threat, especially in the aforementioned countries [3]. The characteristics of olive mill waste vary, depending on factors such as the method of extraction, variety, and maturity of olives, region of origin, climatic conditions, and associated cultivation/processing methods [4]. The disposal of these wastes is very crucial because olive mill waste has been shown to affect the physical and chemical properties of soil and its microbial community, while several studies have evidenced its phytotoxic effects and antimicrobial activity. Olive oil waste water can be toxictoanaerobic bacteria, which may in hibit conventional secondary and anaerobic treatments in municipal treatment plants [5]. Among the possible technologies for recycling the two-phase olive-mill wastewater (TPOMW), composting is one of the most promising options to transform this material into a valuable organic commodity [6]. Composting is a bio-chemical aerobic degradation process of organic waste materials. Under suitable conditions, composting has three consecutive phases: (a) An initial activation phase, (b) a thermophilic phase recognized by a sudden temperature increase, and (c) a mesophilic phase where the organic materials cool down to the surrounding temperature [7]. Microbial metabolic activities generate heat, which leads to physicochemical changes of the organic matter into biomass, CO_2, and humus-like end-products; at the end of the process, a stable, humus-rich, complex mixture is produced [8]. This process leads to the formation of materials that are not associated with the microbiota inhibition and phytotoxicity, which relate to the direct application of the untreated wastes onto the soil [6].

Our group has previously extracted and characterized emulsifiers from food processing waste, e.g., from quince seed [9], winery waste [10], and olive mill waste [11]. In a recent work of ours [12], olive compost has been shown to produce interfacially-active materials with substantial emulsifying capacity. The emulsifiers obtained after composting olive mill waste showed better emulsifying and stabilizing properties compared to the ones produced from uncomposted waste. However, the full composting process cycle is time consuming (6 months). Waste processing is a very low-return operation, so it is crucial to restrict its timescale. As non-composted olive processing waste extracts have some emulsifying capacity [11], it is worthy to investigate whether partial composting is sufficient to increase the emulsifying capacity in a restricted amount of time, e.g., with one month's treatment, rather than the full six-months operation. To this end, this work investigates the capacity of olive processing waste to produce, after a short (1-month) accelerated composting, extracts with enhanced emulsifying capacity.

2. Materials and Methods

2.1. Materials

Tris (hydroxymethyl)-aminomethane, sodium phosphate dihydrate and trihydrate, sodium hydroxide, and acetic acid were purchased from Merck (Darmstand, Germany). Petroleum ether was obtained from Sigma (St. Louis, MO, USA) and hydrochloric acid from Chem-lab NV (Zedelgem, Belgium). Miglyol812 N (Cremer Oleo GmbH & Co., Hamburg, Germany) was used as the oil phase for the interfacial measurements. Extracts were filtered by Whatman filter papers (125mm). Ultrapure water (18.2 MΩ) was obtained from an Ultraclear Ro DI 30 device (Evoqua Lab, Pittsburgh, PA, USA).

2.2. Composting

Olive mill waste (79% *w/w* moisture) on 1.4 m^3 was collected from a two-phase organic olive mill in Kefalonia, Greece. After 2 days of ambient drying, 0.7 m^3 of fresh olive leaves and 10 kg of NH$_4$NO$_3$ fertilizer (Nutrammon, Hellagrolip SA, Athens, Greece) were added with the waste into a purpose-built pit as layers of raw waste, olive leaves, and fertilizer. The compost pile was covered with a plastic sheet to protect it from winter rains. The sheet was removed between storms to promote aeration. The compost core temperature was recorded weekly. Following the beginning of the aerobic fermentation stage, the compost pile was bi-weekly overturned by transfer into an adjacent pit.

2.3. Extraction

The 1-month compost was placed in a vacuum oven (90 °C) until it dried and then it underwent Soxhlet extraction of its lipids with petroleum ether. 10 g of this sample were extracted with either: 0.05 M sodium acetate buffer at pH 5 at 70 °C for 30 min (from now sample OC5), or 0.05 M phosphate buffer at pH 7 at 70 °C for 30 min (from now sample OC7), or 0.05 M tris buffer pH 9 at 70 °C for 30 min (from now sample OC9). The extractions were held in parallel (i.e., each defatted sample was extracted at one pH value (at pH 5, 7, or 9), while extractions at a single pH value were performed sequentially for 3 consecutive times, using 100mL buffer per run. The samples were then centrifuged for 25 min at 3500 rpm in order to separate the extracted matter from the insoluble residue. The supernatants were filtered out and then lyophilized and stored for further use. The materials were reconstituted for use by dissolving in an appropriate volume of de-ionized water under magnetic stirring. They were then enclosed in a dialysis membrane (3500 MWCO) and were immersed in ultrapure water. The water was renewed thrice per day over 2 days. The dialysis membrane content was lyophilized and stored.

2.4. Emulsion Preparation

One of either OC5, OC7, or OC9 (see Section 2.3) were added into buffers of 10 mM Trizma buffer and 1 mM sodium azide set at pH 7 as to prepare 8 mg mL^{-1} solutions. These were magnetically stirred as to dissolve the extracts and were then mixed with miglyol acting as a model oil phase (oil volume fraction $\varphi = 0.1$). These two phases (buffered solution of extracts and miglyol) were magnetically stirred as to prepare a crude pre-mix. This pre-mix was treated with a laboratory ultrasonic homogenizer (Hielscher UP-100H, Teltow, Germany) for a duration of 30 s of continuous treatment. For their long-term monitoring, each emulsion was transferred into a sealed tube and was stored at ambient temperature under dark and quiescent conditions.

2.5. Size Exclusion Chromatography (SEC)

Size exclusion chromatographs were obtained using a setup made of (i) a SpectraSystem SCM 1000 degasser (Thermo Separation Products, San Jose, CA, USA), followed in a series by (ii) a SpectraSystem P 2000 chromatographic pump (Thermo Separation Products, San Jose, CA, USA), then (iii) a column system comprising of a 2 μm frit (Idex, Oak Harbor, WA, USA), then (iv) a GPC/SEC PL-Aquagel-OH 50 × 7.5 mm 8 μm guard column (Varian Inc., Palo Alto, CA, USA), then (v) two GPC/SEC PL-Aquagel-OH 300 × 7.5 mm columns (Varian Inc.) placed in a Model 605 column oven (Scientific Systems Incorporated, State College, PA, USA) set at 30 °C, then (vi) a UV detector recording absorbance at 280 nm (Rigas Labs, Thessaloniki, Greece), and (vii) a BI-MwA laser light scattering detector (MALLS) (Brookhaven Instruments Corporation, Brookhaven, Holtsville, NY, USA). Recording and treating results were handled with ParSec, a dedicated software (ParSec, Brookhaven Instruments Corporation, Brookhaven, Holtsville, NY, USA). Measurements were carried by means of filtering using a 1 μm syringe filter, followed by injecting 200 μL of 8 mg mL^{-1} sample at a flow rate of 0.8 mL min^{-1}. The reader should be reminded here that SEC lets elute the larger-sized molecules first, then followed by molecules of smaller sizes, ending with the smallest of molecules.

2.6. Fourier-Transform Infra-Red Spectroscopy (FTIR)

Fourier transform infra-red (FTIR) spectra of the solid samples were recorded using the Attenuated Total Reflection (ATR) Smart Orbit diamond reflection accessory (Thermo Electron Corporation, Madison, WI, USA) of a Thermo Nicolet 380 IR spectrometer (Thermo Electron Corporation).

2.7. Measurements of Droplet Distribution

The droplet size distributions of miglyol-in-water emulsions emulsified by the materials under study were measured a Malvern Mastersizer 2000 (Malvern Instrument, Malvern, Worcesteshire, UK) apparatus operating with a Hydro MU liquid sampler (Malvern Instrument). The results were treated using the Mie scattering model, assuming spherical particles, a continuous phase diffraction index of 1.33, a dispersed phase refractive index of 1.42, and an absorbance value of 0.1.

2.8. Zeta Potential Measurements

Zeta potential values were obtained with a Brockhaven ZetaPALS apparatus (Brookhaven Instruments Corporation, Brookhaven, Holtsville, NY, USA). Measurements were taken at a temperature of 25 °C in a 10 mM tris buffer set at pH 7, assuming a continuous phase refractive index of 1.33 and a viscosity of 0.89 Pa s. In order to eliminate artefacts due to multiple scattering, all samples were diluted into the buffer and measured again.

2.9. Emulsion Morphology

Micrographs were taken with an inverted Zeiss LSM 700 confocal microscope (Carl Zeiss, CZ Microscopy GmbH, Jena, Germany) in optical mode with a 20× lens. Prior to microscopic examination, 10 μL of 0.1mg mL^{-1} Nile Red and 10 μL of 0.1 mg mL^{-1} Nile Blue were added into each emulsion.

A drop of each emulsion was placed on a welled glass slide and was covered with a coverslip before imaging.

3. Results and Discussion

Figure 1 shows the results of size exclusion chromatography (SEC) for the products obtained by extractions at pH 5, pH 7, and pH 9 (henceforth to be called OC5, OC7, and OC9 respectively) from the initial material (prior to composting). The static light scattering (SLS) detector here records the angular intensity at 90°. At the visible wavelengths used, this signal was sensitive to the molecules of larger sizes (i.e., typically the ones eluting at lower volumes). The second detector (ultraviolet, UV, measuring here the absorbance at 280 nm) was expected to detect proteins based on the absorbance of their amino acids, namely near-UV absorbing Tyr, Trp, and, to a lesser extent, Phe and –S–S– bonds [13].

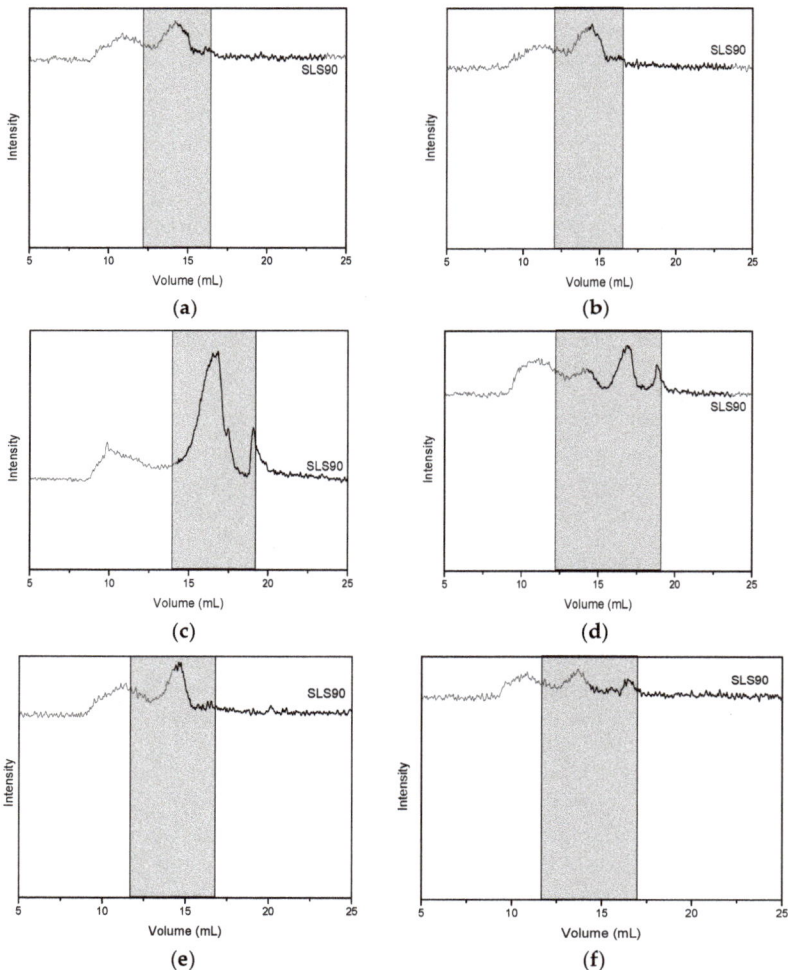

Figure 1. Size exclusion chromatograms of the extracts obtained from non-composted mixtures on the left column (**a,c,e**) and size exclusion chromatograms of the extracts obtained from partially composted mixtures on the right column (**b,d,f**). Top to bottom: Extracts obtained at pH 5 (**a,b**), pH 7 (**c,d**), and pH 9 (**e,f**). The greyed out areas highlight the regions of significant UV absorbance at 280 nm.

In samples OC5 (Figure 1a) and OC7 (Figure 1b), static light scattering (SLS) recorded two populations that scattered light at 90°: One eluting between 10 and 12 mL, and another between 12.5 and 15 mL. The second population absorbed strongly at 280 nm (see the chromatogram of the UV data). In sample OC9 (the one extracted at pH 9, Figure 1c), three scattering populations were distinct: One between 8 and 12 mL, corresponding to the elution times of dextran standards of MW > 1 MDa, a second one 14 and 17 (with a small peak at 17 mL), corresponding to the elution time of dextrans of several tens of kDa, and the third one at 18.5 mL, corresponding to smaller molecules (below 1 kDa). The populations corresponding to 14–17 mL, 17 mL, and 18.5 mL all absorbed at 280 nm. This suggested that the larger populations (those that elute prior to 15 mL) were composed of non-UV absorbing moieties; the only significant candidates for such molecules in plant-based food were polysaccharides. The second peaks should be attributed to proteinic structures, as they absorbed at 280 nm; the same applied for the smaller peak eluting immediately after the larger ones; while the smaller entities eluting at 18.5 mL in OC9 should be the breakdown products of proteins and other UV-absorbing molecules, such as phenolics.

Figure 1 also showed the evolution of these populations after the partial (one month) composting of the material: Samples OC5 showed a remarkable reduction in the size of its components (Figure 1d): The initial population of non-UV absorbing polysaccharides changed its shape from a bimodal peak to a monomodal one, with the entire area shifting to larger elution times. The protein peak, initially between 12.5 and 15 mL (see uncomposted samples in Figure 1), broke down into a smaller peak of the same size, and a much larger one UV-absorbing peak eluting at 17 mL, suggesting the break-up of the protein population during composting. A further new peak appeared at much higher elution times (19 mL), corresponding to the elution time of standard of MW below 1000 Da; that is, this peak was the individual amino acids or oligopeptides deriving from the cleavage of the initial proteinic population. While such changes are not readily apparent in OC7 (Figure 1e) and OC9 (Figure 1f), where the main populations appeared to remain largely unaffected by the short composting process. Overall, OC5 showed promise as the partially-completed composting process started cleaving the macromolecules into smaller ones, of possibly better emulsifying characteristics.

The zeta potential of OC5 was measured to be -9.3 ± 1.0 mV, while that of OC7 was measured to be -26.8 ± 1.0 mV; the reader should be reminded that all zeta potential measurements were taken at pH 7, using samples extracted at different pH values (OC5, OC7, OC9); so their different zeta potentials reflected real compositional differences. That suggested that entities of higher density in ionizable moieties (e.g., carboxyls) were extracted at pH 7 and pH 9 (as compared to pH 5); this could explain the smaller content of degradation products (or products of lower MW) in OC7 and C9 as compared to OC5: For a molecule to be extracted from a plant matrix, the force between the hydrating water and the molecule under extraction should be larger than the forces acting between that molecule and the plant matrix [14]. At pH 7 or 9, the charges were higher than at pH 5 (as at pH 5, most proteins were closer to their pI). So electrostatic interactions between the plant matrix and the molecules under extraction were weaker, facilitating their extraction.

Figure 2 shows the Fourier-transform infra-red (FTIR) spectra of the three materials, as obtained by direct application of the samples on an ATR module. No differences were expected to be found between proteins and their broken down products, as they contained essentially the same vibrating units. However, the FTIR examination could yield data on the chemical identity of the complex mixtures that form these extracts. The major peaks at 1020–1090 cm^{-1} were typical of polysaccharidic entities [15]). The peak groups from 1530 to 1630 cm^{-1} and 1280 to 1450 cm^{-1} were due to proteins and other peptidic entities (i.e., proteins, oligopeptides, and individual aminoacids), attributable to C=O stretching (amide I region) and to N–H bending (amide II region) [16]. Both polysaccharides and proteins have been detected as separate populations in the SEC (Figure 1). A mild shoulder existed at 1540 cm^{-1}, which was more pronounced in samples OC7 and OC9. This was due to carboxylate bending and correlated well with the higher absolute values of (negative) zeta potential of these two samples. The shoulders around 3040 and 3560 cm^{-1} were typical stretching vibrations of O–H units (present in all sugars, peptides, and residual water) [17].

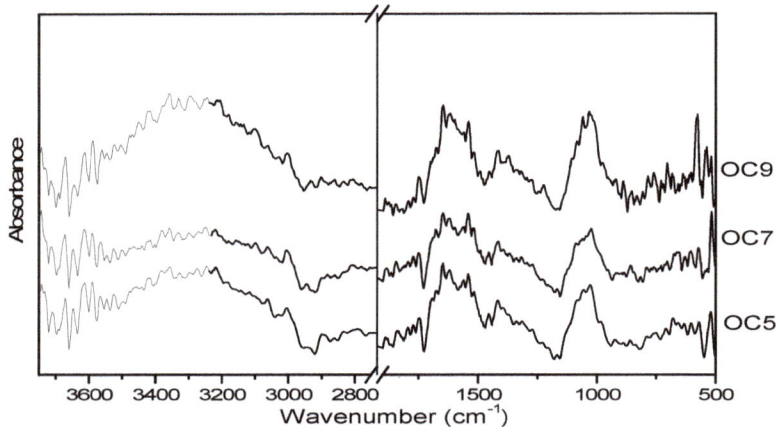

Figure 2. Fourier-transform infra-red spectroscopy (FTIR) spectra of the partially composted waste extracts obtained at pH 5 (OC5), at pH 7 (OC7), and at pH 9 (OC9).

The extract obtained at pH 5 (OC5) has been shown in Figure 1 to comprise of more low-MW components, as the composting process has begun to affect the sample in question. So OC5 has been used as an emulsifier for oil–in–water emulsions set at pH 3 and pH 7 as to simulate neutral and acidic soft foods, respectively. Figure 3 shows the size distributions of the produced droplets and flocs at these two pH values, as they evolved over one week of storage. The top part shows the droplet size distributions for emulsions at pH 3 (Figure 3a). The initial droplets were centered around two populations, around 0.2 and 1 μm, and a population of larger droplets or flocs also existed around 20 μm. These were smaller than the droplets obtained from uncomposted olive extract [11], while they were also smaller than the droplets obtained from emulsifiers from winery extract [10], quince seed [9], and okra-derived emulsifiers [18]. The droplet size distributions remained stable after 24 h, although over 7 days a small 1 μm peak remained, most of the oil existed at entities of 100 μm or above. In that aspect, running the same samples after insertion of 1 g dL^{-1} sodium dodecyl sulphate (SDS) did not change the droplet size distribution (results not shown); that is, displacement of the interfacial layer by SDS does not result in the breaking-up of any flocs; so the larger-sized peaks were most probably individual droplets produced either by coalescence or by Ostwald ripening. The latter mechanism of destabilization has also been clearly demonstrated as the main threat to the stability of emulsions prepared at pH 3 by okra extracts [18]. That suggests that the emulsifiers obtained by OC5 are promising for use in acidic foods, given that the product shelf-life will be restricted.

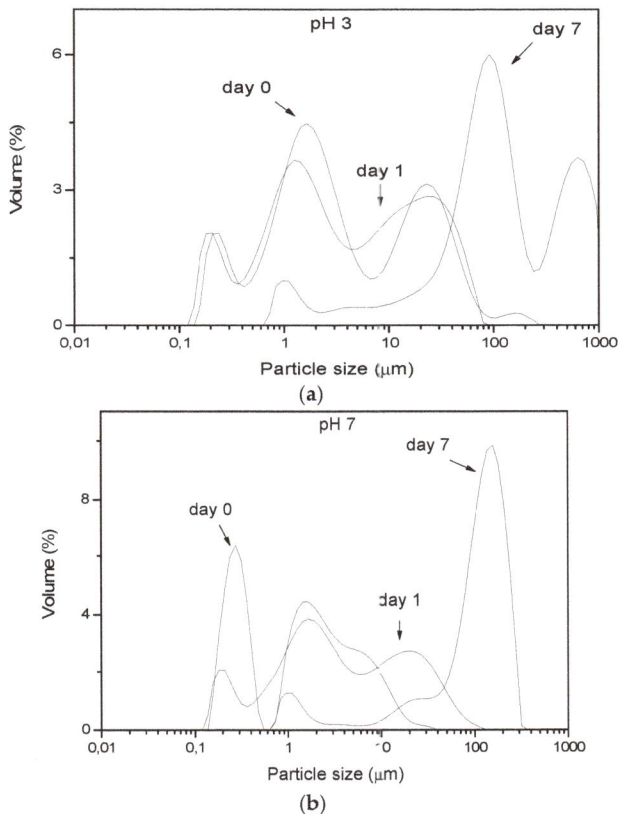

Figure 3. Size distribution of the particles in miglyol-in-water emulsions prepared usingas emulsifier the extract obtained from partially composted olive waste–leaf mixtures extracted at pH 5 (OC5). The emulsions pH is 3 (**a**) and 7 (**b**).

The image obtained at pH 7 (Figure 3b) was different by means of the fast increase of the droplet size within 24 h from preparation. This is reminiscent of the inability of okra extracts to emulsify at pH 7 [18], while non-composted olive mill waste is shown to produce stable emulsions, possibly due to the Pickering action of the larger, non-hydrolyzed macromolecules of the non-composted sample [11]. Micrographic examination of the systems under study (Figure 4) showed clearly the increase in droplet sizes over time; while at pH 3 a few droplets grew at the expense of the smaller ones, at pH 7 the large droplets were dominant, in agreement with the laser particle sizing data of Figure 3. The increase of the larger droplets at the expense of the smaller ones was very clearly depicted as a large number of small droplets flocculated with single large droplets at pH 3 (upper right micrographs, pH 3 at 7 days). Similar observations could also be made for the large droplets at pH 7 (lower part micrographs). A rightward shift of the droplet size distribution, coupled with the growing was the size of the larger droplets at the expense of the smaller ones, was indicative of Ostwald ripening, rather than coalescence [14]. This means that a strong mechanical layer, capable of protecting against droplets merging/coalescing, was indeed formed by the novel emulsifiers; in order to control; the stability of these emulsions, one should seek to control parameters which influence Ostwald ripening rather than coalescence, such as the surface elasticity, the solubility of the oil into water, the surface tension and the viscosity.

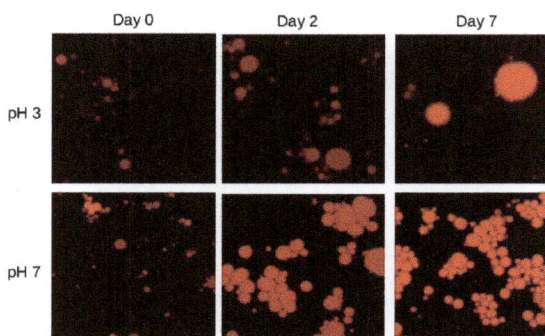

Figure 4. Confocal micrographs of miglyol-in-water emulsions prepared using the extract obtained from partially composted olive waste+leaf mixtures at extraction pH 5 (OC5). The pH values of the emulsions are 3 (top) and 7 (bottom); migrographs are taken (from left to right) upon preparation; after two days of storage; and after seven days of storage.

These observations make an interesting and promising opening for further investigation of the potential of non-complete composting in producing high added-value products. The extraction of hydrocolloids, emulsifiers, or other polyphenolics from the partially-composted agricultural waste can provide a novel, fast, and economic way of valorization of the extensive waste of the agricultural industry. This is an unexplored path that is worth further investigation.

4. Conclusions

Partial composting of olive processing solid wastes can yield emulsifiers that are capable of absorbing onto the oil–water interfaces of acidic emulsions and stabilizing against coalescence; they are less successful in stabilizing against Ostwald ripening. The optimum extraction pH is 5. The ability of these emulsifiers to stabilize the emulsions for some days is due to the presence of the break-down products the solid waste's proteinic and polysaccharidic components that are produced during composting. Extractions at higher pH values do not yield such break-down products, as electrostatic attractions keep the products attached to their plant matrix. Overall, this work serves as a demonstration of the capacity of restricted composting to produce high added-value products.

Author Contributions: Data curation, A.K.M., E.P.S., A.G. and C.R.; investigation, A.K. and V.K.; methodology, C.G., A.K. and C.R.; supervision, C.G. and C.R.; writing—original draft, A.K. and V.K.; writing—review and editing, C.G., E.K., A.K.M., E.P.S., A.G. and C.R.

Funding: "Targeted Actions to Promote Research and Technology in Areas of Regional Specialization and New Competitive Areas in International Level" funded by the Operational Programme "Ionian Islands 2014–2020" and co-financed by Greece and the European Union (Eur: 135.000 euro).

Acknowledgments: We acknowledge the support of this work by the project "Valorization of Olive Mill Waste for the development of high added value products." (MIS 5006879) which is implemented under the Action "Targeted Actions to Promote Research and Technology in Areas of Regional Specialization and New Competitive Areas in International Level" funded by the Operational Programme "Ionian Islands 2014–2020" and co-financed by Greece and the European Union (European Regional Development Fund).

Conflicts of Interest: The authors declare no conflict of interest.

References

1. Muktadirul Bari Chowdhury, A.K.M.; Akratos, C.S.; Vayenas, D.V.; Pavlou, S. Olive mill waste composting: A review. *Int. Biodeter. Biodegr.* **2013**, *85*, 108–119. [CrossRef]
2. Alburquerque, J.A.J.; García, G.D.; Cegarra, J. Effects of bulking agent on the composting of "alperujo", the solid by-product of the two-phase centrifugation method for olive oil extraction. *Process Biochem.* **2006**, *41*, 127–132. [CrossRef]

3. Arvanitoyannis, I.S.; Kassaveti, A. Current and potential uses of composted olive oil waste. *Int. J. Food Sci. Technol.* **2007**, *42*, 281–295. [CrossRef]

4. Ouzounidou, G.Z. Raw and Microbiologically Detoxified Olive Mill Waste and Their Impact on Plant Growth. *Terr. Aquat. Environ. Toxicol.* **2010**, *4*, 21–38.

5. Karaouzas, I.S. Spatial and temporal effects of olive mill wastewaters to stream macroinvertebrates and aquatic ecosystems status. *Water Res.* **2011**, *45*, 6334–6346. [CrossRef] [PubMed]

6. Cayuela, M.P. Duckweed (*Lemna gibba*) growth inhibition bioassay for evaluating the toxicity of olive mill wastes before and during composting. *Chemosphere* **2007**, *63*, 1985–1991. [CrossRef] [PubMed]

7. Ryckeboer, J.; Mergaert, J.; Coosemams, J.; Deprins, K.; Swings, J. Microbiological Aspects of Biowaste during Composting in a Monitored Compost Bin. *J. Appl. Microbiol.* **2003**, *94*, 127–137. [CrossRef] [PubMed]

8. Cooperband, L. *Building Soil Organic Matter with Organic Amendments*; Center for Integrated Agricultural Systems, University of Wisconsin-Madison: Madison, WI, USA, 2002; pp. 1–16.

9. Ritzoulis, C.; Marini, E.; Aslanidou, A.; Georgiadis, N.; Karayannakidis, P.; Koukiotis, C.; Tzimpilis, E. Hydrocolloids from quince seed: Extraction, characterization, and study of their emulsifying/stabilizing capacity. *Food Hydrocoll.* **2014**, *42*, 178–186. [CrossRef]

10. Pavlou, A.; Ritzoulis, C.; Filotheou, A.; Panayiotou, C. Emulsifiers extracted from winery waste. *Waste Biomass Valoriz.* **2016**, *7*, 533–542. [CrossRef]

11. Filotheou, A.; Ritzoulis, C.; Avgidou, M.; Kalogianni, E.; Pavlou, A.; Panayiotou, C. Novel emulsifiers from olive processing solid waste. *Food Hydrocoll.* **2015**, *48*, 274–281. [CrossRef]

12. Koliastasi, A.; Kompothekra, V.; Giotis, C.; Moustakas, A.K.; Scotti, E.P.; Gerakis, A.; Kalogianni, E.P.; Georgiou, D.; Ritzoulis, C. Novel emulsifiers from olive mill compost. *FOODHYD* **2019**. under review.

13. Aitken, A.; Learmoth, P. Protein determination by UV absorption. In *The Protein Protocols Handbook*, 3rd ed.; Walker, J.W., Ed.; Humana Press: Totowua, NJ, USA, 2009; pp. 41–45.

14. Ritzoulis, C. Mucilage formation in food: A review on the example of okra. *Int. J. Food Sci. Technol.* **2017**, *52*, 59–67. [CrossRef]

15. Kacurakova, M.C.; Ebringerova, A. FT-IR study of plant cell wall model compounds: Pectic polysaccharides and hemicelluloses. *Carbohydr. Polym.* **2000**, *43*, 95–203. [CrossRef]

16. Ferreira, A.S.; Nunes, A.; Castro, A.; Ferreira, P.; Coimbra, M.A. Influence of grape pomace extract incorporation on chitosan films properties. *Carbohydr. Polym.* **2014**, *113*, 490–499. [CrossRef] [PubMed]

17. Sun, X.F.; Xu, F.; Zhao, H.; Sun, R.C.; Fowler, P.; Baird, M S. Physicochemical characterisation of residual hemicelluloses isolated with cyanamide-activated hydrogen peroxide from organosolv pre-treated wheat straw. *Bioresour. Technol.* **2005**, *96*, 1342–1349. [CrossRef] [PubMed]

18. Alba, K.; Ritzoulis, C.; Georgiadis, N.; Kontogiorgos, V. Okra extracts as emulsifiers for acidic emulsions. *Food Res. Int.* **2013**, *54*, 1730–1737. [CrossRef]

foods

MDPI

Article

Evaluation of Biomass and Chitin Production of *Morchella* Mushrooms Grown on Starch-Based Substrates

Aikaterini Papadaki [1,2,*], Panagiota Diamantopoulou [2], Seraphim Papanikolaou [1] and Antonios Philippoussis [2,†]

[1] Laboratory of Edible Fungi, Institute of Technology of Agricultural Products (ITAP), Hellenic Agricultural Organization - Demeter, 1 Sofokli Venizelou Street, 14123 -Lykovryssi, 14123 Attiki, Greece
[2] Department of Food Science & Technology, Agricultural University of Athens, Iera Odos 75, 11855 Athens, Greece
* Correspondence: kpapadaki@aua.gr
† Dedicated to his memory.

Received: 12 June 2019; Accepted: 28 June 2019; Published: 1 July 2019

Abstract: *Morchella* sp. is one of the most expensive mushrooms with a high nutritional profile. In this study, the polysaccharide content of *Morchella* species was investigated. Specifically, mycelium growth rate, biomass production, sclerotia formation, and glucosamine and total polysaccharides content of six *Morchella* species grown on a starch-based media were evaluated. Submerged fermentations in potato dextrose broth resulted in a glucosamine content of around 3.0%. In solid-state fermentations (SSF), using potato dextrose agar, a high linear growth rate (20.6 mm/day) was determined. Increased glucosamine and total polysaccharides content were observed after the formation of sclerotia. Biomass and glucosamine content were correlated, and the equations were used for the indirect estimation of biomass in SSF with agro-industrial starch-based materials. Wheat grains (WG), potato peels (PP), and a mixture of 1:1 of them (WG–PP) were evaluated as substrates. Results showed that the highest growth rate of 9.05 mm/day was determined on WG and the maximum biomass yield (407 mg/g) on WG–PP. The total polysaccharide content reached up to 18.4% of dried biomass in WG–PP. The results of the present study proved encouraging for the efficient bioconversion of potato and other starch-based agro-industrial waste streams to morel biomass and sclerotia eliciting nutritional and bioactive value.

Keywords: *Morchella*; morel mushrooms; bioprocess development; solid state fermentation;.food processing; glucosamine; polysaccharides; bioactive compounds

1. Introduction

Mushrooms are widely known for their taste and flavor presenting many functional properties, primarily due to their unique chemical composition. They are consumed either fresh or processed. For instance, mushroom powder is used as a food additive to increase the content of dietary fibers in foods or as a partial flour substitute in bakery products [1]. In addition to fresh or dried mushrooms, fungal mycelium is also a rich source of bioactive compounds with many functional properties and has been suggested as an alternative mushroom product for human consumption [2]. Agro-industrial wastes and side streams have been converted into various bioactive compounds, including polysaccharides and enzymes, through mushroom cultivation [3–5]. Mushroom-derived polysaccharides, such as glucans and chitin, have attracted research interest mainly due to their antioxidant, anti-inflammatory, antitumor, and immune-stimulation activity [6].

Chitin is a structural polysaccharide of the fungal cell wall, composed of $\beta(1,4)$-linked units of N-acetyl-d-glucosamine [7]. The chitin content of fungal mycelium can reach up to 42% (w/w, dry mass)

depending on the fungal species, fermentation mode (submerged or solid-state fermentation), and the substrate [7–9]. Chitin and its deacetylated derivate, chitosan, are valuable compounds finding many applications, namely in food and pharmaceutical industry, due to their antimicrobial and antioxidant properties [7]. They have been used as functional food components, edible films [7], immobilization support material for enzymes [10], delivery system for food ingredients, such as carotenoids [11], and for the treatment of human diseases [12]. Chitin production reached around 28,000 tons in 2015 [13], mainly produced from by-products of the seafood industry [7,14]. The chemical extraction of glucosamine from seafood by-products requires high energy consumption and the use of strong chemical reagents. However, the main disadvantages of using seafood by-products, as glucosamine sources, are the seasonal and geographical limitations, the potential allergic effects of the product, and the formation of products with inconsistent composition and physicochemical properties [8,15]. Thus, chitin production by fungal fermentations has been suggested as an alternative source, as glucosamine can be produced in controlled conditions by simpler extraction process and using less-aggressive chemicals [8,14].

Morels (*Morchella* spp.), belonging to the Helvellaceae family of ascomycetes, are among the most desirable edible wild mushrooms. Researchers have suggested that the differences in the appearance of morels are due to environmental influences and that the mycelial mass determines the appearance or phenotype of morels in terms of color (yellow or black), yet not the shape. They have attracted research interest due to their commercial value, medicinal properties, and unique taste and flavor [16]. The artificial commercial cultivation of the fruit bodies of morels on various agro-industrial substrates is a difficult process, which is linked to the formation of a heterokaryotic sclerotium [17]. Published studies on the considerable morphological variations in the physiology of *Morchella* sp. in different carbon [18] and nitrogen sources [19] are available, however reports dealing with the production of high-added-value compounds by *Morchella* sp. are scarce. The most recent studies have demonstrated that polysaccharides from *Morchella* sp. present antioxidant [20,21] and antitumor [22] properties. Additionally, sclerotia from mushrooms are considered as an exceptional source of bioactive components characterized for their functional and medicinal properties, such as antitumor and anti-inflammatory activities. Lau and Abdullah [23] pointed out that the biological activities of the sclerotia seemed to be comparable to those of the mycelia. However, their formation and chemical composition are strongly dependent on the substrate and the fermentation conditions [23].

Many researchers have studied those characteristics of *Morchella* mushrooms in submerged fermentations (SmF) [24–26], still little is known about their behavior in solid-state fermentations (SSF) [27]. The aim of the present study was to determine the glucosamine and polysaccharide contents in the mycelia of different *Morchella* strains grown on starch-based substrates. Initially, six different *Morchella* strains were cultivated on commercial starch-based substrates. Biomass production, sclerotia formation, and glucosamine and polysaccharide contents were determined in SmF and SSF. Equations relating biomass production and glucosamine content were established and subsequently applied to estimate the biomass production during SSF on agro-industrial starch-based substrates. To the best of our knowledge, this is the first study reporting the relation between the biomass production and the glucosamine content of *Morchella* strains.

2. Materials and Methods

2.1. Raw Materials

Wheat grains (WG) and wheat bran were purchased from the local market (Athens, Greece). Potato peels (PP), HERMES variety, were kindly provided by the potato processing industry Tasty Foods S.A. (Athens, Greece).

2.2. Morchella Strains, Media, and Culture Conditions

Experiments were carried out using six *Morchella* strains, belonging to the group of yellow and black morels (Table 1). All strains were obtained from the fungal AMRL (Athens Mushroom Research Laboratory) culture collection of the Laboratory of Edible Fungi, Institute of Technology of Agricultural Products (LEF, ITAP). *Morchella* strains were grown on potato dextrose agar (PDA; Merck, Darmstadt, Germany) plates at 26 ± 0.5 °C for seven days. PDA plates were maintained at 4 ± 0.5 °C and used as fermentation inoculums.

Table 1. *Morchella* strains used in the study.

Morels Group	*Morchella* Strain
Yellow morels	*M. rotunda* AMRL 14
	M. vulgaris AMRL 36
	M. crassipes AMRL 52
Black morels	*M. elata* AMRL 63
	M. conica AMRL 74
	M. angusticeps AMRL 82

AMRL: Athens Mushroom Research Laboratory.

2.3. Submerged Fermentations on Commercial Substrates

SmF fermentations were initially conducted for the evaluation of biomass production in commercial starch- and glucose-based media. Potato dextrose broth (PDB) media was prepared by enriching the extract from 300 g/L potatoes with glucose 20 g/L and $CaCO_3$ 2 g/L. In addition, a glucose-based media (GPYB; Glucose Peptone Yeast Broth) was also prepared consisting of: Glucose, 30 g/L; peptone, 3.5 g/L; yeast extract, 2.5 g/L; $CaCO_3$, 2 g/L; KH_2PO_4, 1 g/L; $MgSO_4·7H_2O$, 0.5 g/L; $CaCl_2·2H_2O$, 0.3 g/L; $MnSO_4·H_2O$, 0.04 g/L; $ZnSO_4·7H_2O$, 0.02 g/L; and $FeCl_3·6H_2O$, 0.08 g/L. Both media were used for the SmF cultures in static conditions. Erlenmeyer flasks of 100 mL capacity, containing 20 mL of each liquid medium, were autoclaved for 20 min at 121 ± 0.5 °C, and subsequently inoculated with two agar disks of 6 mm diameter. Inoculum disks were cut from a seven-day-old growing colony on a PDA Petri dish. Static cultures were incubated at 26 ± 0.5 °C for 21 days. Triplicates were made for every sampling to determine biomass production, mycelium glucosamine content, polysaccharide content, and sclerotia formation. The experimental data were fitted by ORIGIN software (OriginPro 8, Originlab Corporation, Northampton, MA, USA).

2.4. Solid-State Fermentations on Commercial Substrates

SSF using PDA plates were employed for the evaluation of biomass production, radius growth rate, and glucosamine content of *Morchella* strains. Also, the cellophane technique was applied for the estimation of the dry weight of the fungal biomass, since it prevents penetration of hyphae into solid medium and makes the separation of the fungus possible [28,29]. Specifically, 20 mL of PDA, prepared as previously described and solidified by the addition of 20 g/L agar, was poured into Petri dishes (90 mm diameter), covered by polyethylene terephthalate (PET) membrane disks. PET membranes were boiled twice for 15 min in deionized water to remove plasticizers [30] and sterilized at 121 ± 0.5 °C for 20 min. Petri dishes were inoculated at the center with a 6 mm diameter mycelium plug and incubated at 26 ± 0.5 °C. At least three replicates per treatment and sampling were used to study the growth of tested strains regarding radius growth rate and sclerotia formation, biomass production, and its glucosamine and polysaccharide content. The experimental data were fitted by ORIGIN software (Northampton, MA, USA).

2.5. Solid-State Fermentations on Agro-Industrial Substrates

Wheat grains (WG), potato peels (PP), and a mixture of them (WG:PP, 1:1) were used for the solid-state cultivation of selected *Morchella* strains. Substrate content was 95% of WG, PP, or WG–PP, supplemented with 5% of wheat bran. PP and WG were washed to remove any wasteful materials. WG were boiled for 20 min and left to cool down. After drainage they were mixed with the previously moistened wheat bran to obtain ~65%–70% moisture content, while the pH ranged from 6.5 to 6.9 after addition of 0.2% (w/w) CaCO₃. Petri dishes (150 mm diameter) were filled with the substrates and autoclaved for 20 min at 121 ± 0.5 °C. Inoculation was carried out with a 9 mm diameter agar disk and incubated at 26 ± 0.5 °C for 30 days. Mycelial growth rate as well as sclerotia formation and maturation was recorded daily. The mycelium concentration in the substrate was indirectly estimated through the regression equations of glucosamine vs. biomass, defined previously in the PDA–PET experiment.

2.6. Analytical Methods

2.6.1. Mycelium Growth Rate

The radius growth rate (Kr) of mycelium (expressed in mm/day), was determined by fitting the growth parameters using the equation [31,32]:

$$r = Kr \cdot t + r_0,$$

where r and r_0 are the colony radius at time t and t_0, respectively, and Kr is the constant growth rate. Measurements of colony diameter on the surface of SSF were taken in two perpendicular directions every 12 or 24 h, until the colony completely covered the Petri dish.

2.6.2. Determination of Biomass and Sugar Consumption in Submerged Fermentations

Samples were withdrawn from SmF at specific intervals for the determination of biomass production and sugars consumption. Fungal biomass was separated from the culture broth by filtration (Whatman No. 2, Buckinghamshire, UK), washed twice with deionized water, and dried at 60 ± 0.5 °C until constant weight. The clear broth was used for the determination of reduced sugars by the 3,5-dinitro-2-hydroxy-benzoic acid (DNS) method [33] and total sugars content was estimated by the phenol–sulfuric acid method according to Dubois et al. [34].

2.6.3. Determination of Biomass in Solid-State Fermentations

The glucosamine present in the fungal cell wall was used to monitor fungal biomass in SSF. Initially, a glucosamine standard curve (glucosamine vs. absorbance) was obtained using various concentrations of N-acetyl-D-glucosamine (Sigma-Aldrich, St. Louis, MO, USA). Subsequently, regression equations of the glucosamine content of each *Morchella* strain were determined using dry biomass (biomass vs. glucosamine). The chitin content of dried biomass was hydrolyzed into N-acetylglucosamine according to the method described by Scotti et al. [35]. Specifically, around 2 g of dry sample was mixed with 5 mL of 72% H₂SO₄ (Merck, Germany) followed by agitation (130 rpm) (rotary shaker, MPM Instruments Srl., M301-OR, Italy) for 30 min at room temperature. Then, samples were diluted with 54 mL deionized water and treated at 121 ± 0.5 °C for 2 h. The hydrolyzate was neutralized (pH 7.0) with a 10 N NaOH solution (Merck, Darmstadt, Germany) and further treated for the quantification of glucosamine [4].

The colorimetric method of Ride and Drysdale [9] was carried out for the determination of glucosamine content. An aliquot of 3 mL of hydrolyzate was obtained and an equal volume of 5% (w/v) NaNO₂ (Merck, Germany) and 5% (w/v) KHSO₄ (Merck, Germany) were added. The mixture was agitated for 15 min and then centrifuged (1500 × g, 2 °C, 2 min) (Hettich Micro22R, Tuttlingen, Germany). Then, 3 mL of the supernatant was obtained followed by addition of 1 mL of 12.5% (w/v) NH₄SO₃NH₂ (Merck, Germany) and agitation for 5 min. Subsequently, 1 mL of freshly prepared 0.5%

(w/v) 3-methyl-2-benzothiazolone hydrazone hydrochloride (MBTH; Sigma-Aldrich, St. Louis, MO, USA) was added, followed by heating in a boiling water bath for 3 min. The mixtures were cooled down and 1 mL of freshly prepared 0.5% (w/v) $FeCl_3$ (Alfa Aesar, Kandel, Germany) was added to each sample. After standing for 30 min, the solution was centrifuged and the absorbance was read at 650 nm (Jasco V-530 UV/VIS spectrophotometer, Jasco, Tokyo, Japan). The same protocol (hydrolysis and analysis) was followed using the unfermented medium as a blank. In SSF experiment results were expressed as mg fungal biomass per g of dry substrate.

2.6.4. Determination of Total Polysaccharide Content

The content of polysaccharides was determined following a modification of the anthrone method [36–38] using sucrose as a standard. Specifically, anthrone reagent (0.2 g/L) was prepared in an aqueous solution of H_2SO_4 (70%, w/v), then the mixture was boiled for 15 min and rapidly cooled. The reagent was kept in a dark and cool (4 °C) place for 24 h. For the determination of total polysaccharides (TP), the dried biomass was extracted with 10 mL of H_2SO_4 70% for 30 min using an agitation rate of 130 rpm. Subsequently, 5 mL anthrone reagent was added to 1 mL of the extracted sample. The mixture was first cooled in water, then swirled, heated in a water bath at 60 ± 0.5 °C for 8 min, and rapidly cooled. The absorbance of samples was measured at 630 nm (Jasco, V-530 UV/VIS, Tokyo, Japan) after 1 h.

3. Results and Discussion

3.1. Submerged Fermentations

3.1.1. Biomass Production, Glucosamine, and Total Polysaccharide Content

The quantification of fungal biomass is unfeasible in SSF due to the penetration of the hyphae into the solid substrate [8,39]. Among the different methods for monitoring fungal biomass in SSF, the estimation of glucosamine content is considered a representative indicator [38]. Therefore, SmF fermentations were initially carried out in PDB and GPYB to determine the relationship between glucosamine content and dry biomass.

Morchella strains were cultivated for 21 days in PDB and GPYB. In all cases, *Morchella* strains were able to consume more than 90% of the initial sugar concentration, except for AMRL 82 in GPYB (75%). The highest biomass production and the respective glucosamine and polysaccharide contents are shown in Table 2. Biomass production ranged from 9.3 to 11.1 g/L for yellow morels and from 9.4 to 14.2 g/L for the black ones. The glucosamine content varied between 2.3%–3.7% (*w/w*) for both yellow and black morels. Although, in most cases, higher biomass concentrations were observed in the GPYB medium (~10–14 g/L), the highest biomass productivity was attained in PDB for almost all strains (~1.3–1.7 g/L/day). The yield of biomass (Yx/s) based on the utilized substrate was calculated in order to establish the relationship between microbial growth and substrate consumption (Table 2). Biomass yield ranged around 0.41–0.43 g/g in all cases, except for AMRL 36 and AMRL 82, which was above 0.5 g/g in GPYB medium. This probably means that the starch-based substrate promoted biomass productivity. Previous studies highlighted that starch has been characterized as a superior carbon source for *Morchella* strains [40,41]. Zhang et al. [42] reported the ability of *Morchella esculenta* to degrade starch and upgrade the nutritional value of cornmeal during SSF; it was attributed to the high α-amylase production (215 U/g on the 20th cultivation day). Xing et al. [43] reported high biomass production (12.6 g/L) by the black morel *Morchella conica* grown on a synthetic sucrose-based medium. Other studies have reported lower concentrations, ranging from 2.6 to 10 g/L for both black and yellow morels using various carbon sources [25,40,44–47].

Table 2. Maximum biomass production and glucosamine and total polysaccharide (TP) content of *Morchella* strains, during submerged fermentations in potato dextrose broth (PDB) and glucose-based broth (GPYB).

Morchella Strains	Medium	Time (days)	Biomass X (g/L)	Biomass Yield Yx/s (g/g)	Productivity P_X (g/L/day)	Glucosamine (%, w/w)	TP (%, w/w)
AMRL 14	PDB	14	10.7 ± 0.67	0.42	0.76	3.0 ± 0.07	9.6 ± 0.60
	GPYB	14	11.1 ± 0.80	0.41	0.79	3.7 ± 0.08	10.6 ± 0.03
AMRL 36	PDB	7	9.8 ± 0.45	0.41	1.40	2.5 ± 0.07	9.7 ± 0.02
	GPYB	7	10.2 ± 0.07	0.52	1.46	2.3 ± 0.08	10.8 ± 0.31
AMRL 52	PDB	7	9.3 ± 0.40	0.43	1.33	2.5 ± 0.09	10.1 ± 0.05
	GPYB	14	10.9 ± 0.35	0.43	0.78	3.1 ± 0.06	11.6 ± 0.30
AMRL 63	PDB	7	10.6 ± 0.45	0.43	1.51	2.3 ± 0.06	11.8 ± 0.12
	GPYB	14	14.2 ± 0.10	0.47	1.01	3.0 ± 0.04	12.2 ± 0.06
AMRL 74	PDB	7	11.8 ± 0.75	0.47	1.69	2.4 ± 0.04	10.3 ± 0.14
	GPYB	14	9.4 ± 0.97	0.32	0.67	3.1 ± 0.06	10.5 ± 0.10
AMRL 82	PDB	14	10.9 ± 0.55	0.42	0.78	3.5 ± 0.07	10.4 ± 0.30
	GPYB	14	12.4 ± 0.05	0.56	0.89	2.8 ± 0.05	10.9 ± 0.23

In this study, glucosamine content varied among *Morchella* strains, from 2.3% to 3.7% of fungal dry matter. Obviously, glucosamine content was dependent on the fungal strain and the composition of the substrate, which has been previously reported by other studies [9,38,39,48]. The chitin content of *Morchella* sp. has been reported to be around 16% [49], but the relation of fungal biomass with glucosamine content on agro-industrial substrates has not been reported.

The polysaccharides present in mycelium and fruiting bodies are classified as glycoconjugates and can be quantified by the anthrone method [50]. Table 2 presents the results of the TP content of *Morchella* mycelia grown in PDB and GPYB. The TP content of mushrooms was influenced by the strains; yellow morels had ~10.4% and black ~11.2 % w/w. The morels AMRL 36 and 63 presented the highest TP content (10.1%–12.2%) among all strains. Morel mushrooms are well known for their rich nutritional composition and, specifically, their sugar profile comprises mainly mannose (up to 43% w/w on a dry basis) and mannitol (up to 11.5%, w/w), followed by glucose, trehalose, fructose, and arabitol [51]. Dried biomass from different culture days was obtained for each strain and their glucosamine content was estimated. The results were correlated and the linear regression equations of glucosamine versus biomass are shown in Table 3. Results demonstrate that biomass production and glucosamine content were found to be highly correlated ($R^2 > 0.97$). The aim was to use the equations to convert glucosamine content into mycelia biomass in the following SSF experiments. This approach has been already successfully applied in SSF of the medicinal mushroom *Lentinula edodes* [4].

Table 3. Linear regression equations of glucosamine (mg) and mycelial biomass (g) of *Morchella* strains grown on potato dextrose broth (PDB) and glucose-based broth (GPYB).

Morchella Strains	Medium	Biomass (y)/Glucosamine (x)	Glucosamine (y)/Biomass (x)	R^2
AMRL 14	PDB	y = 0.0335x − 0.0256	y = 29.499x + 0.8482	0.99
	GPYB	y = 0.0277x − 0.0046	y = 34.845x + 0.4457	0.97
AMRL 36	PDB	y = 0.0376x − 0.0064	y = 26.476x + 0.1888	0.99
	GPYB	y = 0.0336x − 0.0055	y = 29.308x + 0.2457	0.98
AMRL 52	PDB	y = 0.0369x − 0.0026	y = 26.151x + 0.2788	0.97
	GPYB	y = 0.0301x − 0.0002	y = 32.581x + 0.1026	0.98
AMRL 63	PDB	y = 0.0378x + 0.0020	y = 25.753x + 0.0513	0.97
	GPYB	y = 0.0328x − 0.0010	y = 30.057x + 0.0825	0.99
AMRL 74	PDB	y = 0.0385x + 0.0045	y = 25.627x - 0.0563	0.99
	GPYB	y = 0.0306x − 0.0124	y = 32.223x + 0.4669	0.99
AMRL 82	PDB	y = 0.027x − 0.0073	y = 36.685x + 0.3512	0.99
	GPYB	y = 0.0355x − 0.0219	y = 27.988x + 0.6530	0.99

3.2. Solid-State Fermentations on Commercial Substrates

PDA was selected for the evaluation of growth rate, biomass production, glucosamine content, and sclerotia formation in SSF, due to the higher biomass productivity of *Morchella* strains in SmF (Table 2). Morel strains AMRL 14 and AMRL 82 strains were excluded from SSF due to the lower biomass productivity observed in SmF (Table 2).

3.2.1. Mycelial Growth Rate

The mycelial growth rate of four *Morchella* strains was evaluated on PDA medium, with and without the addition of a PET membrane (Figure 1). In the case of SSF without PET membrane, strains from the black morels complex (AMRL 63 and 74) proved to grow faster than yellow morels. Among the black morels, the AMRL 74 strain presented the maximum growth rate (Kr) yielding 22.2 mm/day, whereas AMRL 52 presented the greater growth rate (13.1 mm/day) from the yellow morels complex. Brock [40] determined a growth rate of 21 mm/day for *M. esculenta* in SSF with glucose, as a carbon source. Winder [18] reported lower growth rates for the black morel strains compared to this study. More specifically, the growth rate was around 6 mm/day in SSF with sucrose and mannose as the substrate and 1 mm/day in PDA. Lower growth rates have been also determined for other mushrooms. More specifically, growth rates up to 6.6 mm/day for *Pleurotus* sp., 4.4 mm/day for *L. edodes*, and 18.8 mm/day for *Volvariella volvacea* have been reported during SSF on PDA [52]. In the case of SSF with PDA–PET, the growth rate was remarkably suppressed. The growth rate was reduced by 64%–82%, depending on the *Morchella* strain. The highest growth rate in PDA–PET was 5.7 mm/day from AMRL 63. *Morchella* strains have not been studied in SSF with the presence of the membrane. The only published study using PDA covered with a membrane reports a growth rate of 4.8 mm/day for *Morchella* [53]. Reeslev and Kjøller [54] reported that the presence of the membrane reduced the growth rate of the ascomycota *Paecilomyces farinosus* only by 8%. It could be assumed that the type of membrane can affect the growth rate. For instance, in this study a plastic membrane was employed, whereas Reeslev and Kjøller [54] used a cellulosic membrane.

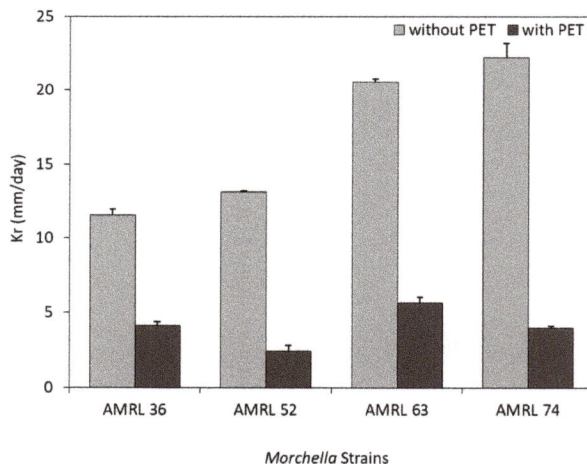

Figure 1. Growth rate of yellow (AMRL 36, 52) and black (AMRL 63, 74) morel strains during solid-state fermentations on potato dextrose agar (PDA), with and without polyethylene terephthalate (PET) membrane.

3.2.2. Sclerotia Formation

Sclerotia formation and maturation was studied on PDA substrate, with and without a PET membrane (Table 4). Maturation was expressed according to the size of sclerotia (immature ≤ 1 mm, mature > 1 mm), whereas the number of sclerotia was classified as follows: Few < 20, adequate > 20, and many > 50. In the case of PDA medium, morel strains AMRL 36, AMRL 52, and AMRL 74 produced few immature sclerotia at the eighth day of cultivation, whereas few mature sclerotia were formed only in the later fermentation stage (21st day). AMRL 63 exhibited more immature sclerotia compared with all the other strains. Immature and mature sclerotia of AMRL 63 appeared earlier.

Table 4. Sclerotia formation and maturation of yellow (AMRL 36, 52) and black (AMRL 63, 74) morel strains grown on PDA medium, with and without polyethylene terephthalate (PET) membrane.

Medium	Time (days)	Strains/Sclerotia Number and Maturation							
		AMRL 36		AMRL 52		AMRL 63		AMRL 74	
		I [1]	M [2]	I	M	I	M	I	M
PDA	8	+	−	+	−	+	−	+	−
	13	+	−	+	−	++	*	+	−
	21	+	*	+	*	++	*	+	*
PDA–PET	12	−	*	−	*	+	*	+	−
	22	+	*	+	**	++	**	+	*
	34	++	*	++	**	++	***	++	*

[1] I: Immature sclerotia (≤ 1 mm); − no sclerotia, + few (< 20), ++ adequate (> 20), +++ many (> 50). [2] M: Mature sclerotia (>1 mm); − no sclerotia, * few (< 20), ** adequate (> 20), *** many (> 50).

Sclerotia formation was significantly impaired by the PET membrane. More matured sclerotia were formed in the case of AMRL 52 and AMRL 63 strains, as compared with PDA without PET. The number of immature and mature sclerotia was also promoted in all strains. Generally, starch has been found to promote sclerotia production [41]. The presence of PET induced a nutritional stress, which in turn promoted sclerotia formation [41]. The present study demonstrated that sclerotia formation was influenced by fermentation time, species, and culture conditions. Similar conclusions have also been mentioned by [41] for *Morchella* sp. Sclerotia are composed of large cells with thick walls and their presence is considered a precursor of fruiting body formation [41]. Additionally, their significance is based on their chemical composition as they are a rich source of bioactive compounds.

3.2.3. Biomass Production, Glucosamine, and Total Polysaccharide Content

The use of membranes as a separation method of mushroom biomass from solid substrates, allows the direct estimation of biomass, hence this method has been widely reported in literature [28]. In our study, PDA plates covered with PET were utilized and the biomass accumulated on the membrane surface was determined. Black morels achieved higher biomass concentrations than yellow morels on the PDA–PET substrate. In particular, the maximum biomass concentration was obtained for AMRL 63 (7.38 g/L) and AMRL 52 (6.47 g/L), both on the 34th day (Figure 2). These strains also presented the highest biomass concentration in PDB fermentation (Section 3.1.1, Table 2). Similar studies with *Morchella* strains have not been published, however the mycorrhizal fungi *Suillus collinitus* and *Pisolithus tinctorius* produced 7.1 and 5.4 g/L of biomass, respectively, in PDA with a membrane [29]. The glucosamine content was higher than 2.5% (w/w) for all *Morchella* strains (Figure 2). A positive correlation was found between biomass production and glucosamine content for all strains ($R^2 > 0.97$ and $R^2 = 0.8$ for AMRL 63). Similar glucosamine contents were obtained in SmF using PDB.

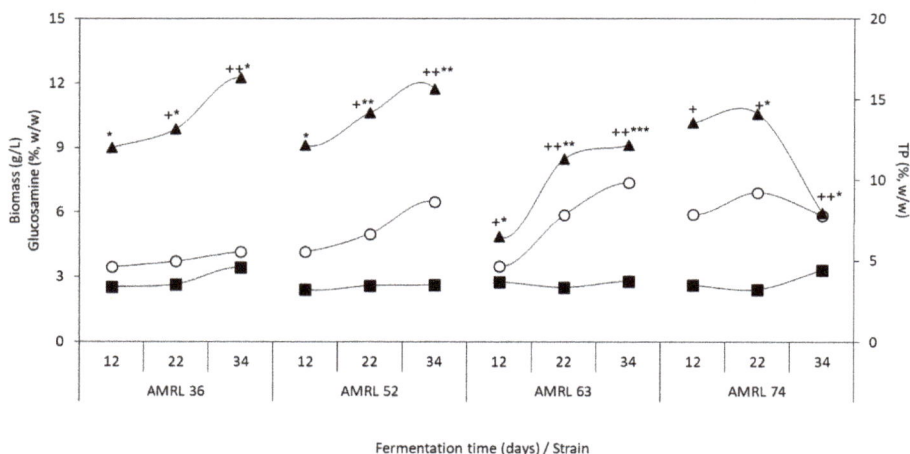

Figure 2. Biomass (○) and sclerotia production (+/*), glucosamine (■) and total polysaccharides (TP) (▲) content of yellow (AMRL 36, 52) and black (AMRL 63, 74) morel strains during solid-state fermentation (12th, 22nd, and 34th day) on potato dextrose agar (PDA), covered by polyethylene terephthalate (PET) membrane. The symbols +/* represents the number of immature/mature sclerotia, respectively (+/*; < 20, ++/**; > 20, +++/***; > 50).

The age of the mycelium affected the TP content (Figure 2), which was found to be more than 15% for the yellow morel strains. The comparison with SmF in PDB revealed that TP content was enhanced in SSF. A positive correlation ($R^2 > 0.94$) was found, between biomass production and TP content, for all strains (for AMRL 74 a positive correlation was found until the 22nd day). This is in agreement with previous results for other mushrooms [55]. On the top of that, Desgranges et al. [38] mentioned that TP content is increased as the age of the mycelium increases for the ascomycota *Beauveria bassiana*. Figure 2 showed that the TP content was influenced by sclerotia formation. In particular, increased TP content was observed as the number of mature and immature sclerotia increased. The determination of higher glucosamine and polysaccharide contents along with the appearance and maturation of sclerotia indicates that sclerotia are rich in chitin and polysaccharides. This has not been mentioned before for *Morchella* sp., however the presence of chitin and β-glucans was identified in sclerotia of *Pleurotus tuber-regium* [56]. The composition of the sclerotia of *Morchella* sp. should be further studied to identify the bioactive compounds and their biological activities.

The dried biomass was collected and correlated with glucosamine content. The linear regression equations were found to be the same, for each strain, as those indicated in Table 2 (PDB medium). These equations were further implemented for the determination of biomass formation in SSF with natural starch-based media.

3.3. Solid-State Fermentations on Agro-Industrial Substrates

SSF in WG, PP, and WG–PP were carried out using the morel strains AMRL 52 and AMRL 63, as they yielded higher biomass, TP content, and sclerotia formation in commercial substrates.

3.3.1. Mycelial Growth Rate

The growth rate of *Morchella* strains AMRL 63 and AMRL 52 was studied on WG, PP, and WG–PP substrates, as depicted in Figure 3. Both *Morchella* strains presented similar growth behavior in all substrates, with the black morel strain exhibiting faster growth rate than the yellow morel strain. Specifically, the highest growth rate of AMRL 63 was detected on WG substrate (9.0 mm/day) and WG–PP (8.8 mm/day), whereas AMRL 52 presented similar growth rates in all substrates (PP, 6.1 mm/day; WG, 5.9 mm/day; WG–PP, 5.6 mm/day).

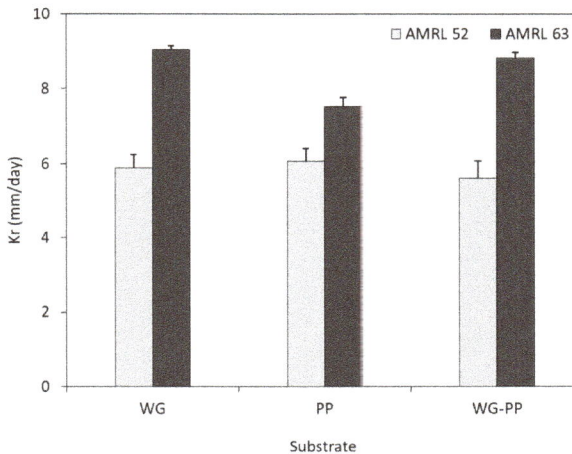

Figure 3. Growth rate of yellow (AMRL 52) and black (AMRL 63) morel strains during solid-state fermentation on wheat grains (WG), potato peels (PP), and a mixture of them (WG–PP, 1:1).

The comparison of PDA and PP, WG, and WG–PP substrates demonstrated that AMRL 63 and AMRL 52 strains exhibited higher extension rates on agro-industrial substrates. This shows that starch-based substrates could be an alternative for the production of *Morchella* mycelium.

The results clearly indicate that the substrate had a crucial role in the growth of *Morchella* strains. For instance, Alvarado-Castillo et al. [57] mentioned that growth rate was significantly affected by the type of grains used in SSF employed in jars. *Morchella* strains presented the highest mycelia growth (more than 30 cm^2) in rye grains, followed by oats, wheat, and maize grains [57]. SSF of other mushrooms, such as *Pleurotus* sp., *L. edodes*, *Ganoderma* sp., and *V. volvacea* among others, have shown that the growth rate is highly dependent on the strain and the substrate [3,52]. *Pleurotus* sp., *Ganoderma* sp., and *Lentinula* sp. presented maximum growth rates ranging from 4.4 to 9.8 mm/day when cultivated on spent mushroom substrate, wheat straw, corn cobs, oak sawdust, and peanut shells) [3,52], whereas growth rate of *V. volvacea* reached 12.5 mm/day in wheat straw [52].

3.3.2. Sclerotia Formation

Different outcomes were obtained in SSF of AMRL 52 and AMRL 63 strains in the WG, PP, and WG–PP substrates (Table 5), as compared to PDA substrate. The quantity of mature sclerotia was lower compared to the PDA–PET medium. *Morchella* strains formed sclerotia in WG and WG–PP substrate, whereas no sclerotia were observed in PP substrate. Among strains, AMRL 63 produced more mature and immature sclerotia on 30th day in WG–PP. It seems therefore that the mycelium growth rate is positively correlated to the sclerotia number. However, Alvarado-Castillo et al. [57] observed that growth rate was inversely related to sclerotia formation in SSF using various grains. Generally, the sclerotia formation of *Morchella* mushrooms is promoted in starch-based substrates [17], due to their content of rapidly metabolized sugars, such as starch and simple sugars. This has been also observed in SSF of *M. esculenta* using a co-substrate of wheat bran and corn starch [58].

Table 5. Sclerotia formation and maturation of yellow AMRL 52 and black AMRL 63 morel strains grown on wheat grains (WG), potato peels (PP), and a mixture of them (WG–PP, 1:1).

Medium	Time (days)	Strains/Sclerotia Number and Maturation			
		AMRL 52		AMRL 63	
		I [1]	M [2]	I	M
WG–PET	20	+	−	+	−
	30	++	−	+	*
PP–PET	20	−	−	−	−
	30	−	−	−	−
WG:PP–PET	20	+	*	++	*
	30	+	*	++	**

[1] I: Immature sclerotia (≤ 1 mm); − no sclerotia, + few (< 20), ++ adequate (> 20), +++ many (> 50). [2] M: Mature sclerotia (>1 mm); − no sclerotia, * few (< 20), ** adequate (> 20), *** many (> 50).

3.3.3. Biomass Production and Total Polysaccharide Content

An indirect estimation of biomass production was applied in SSF using starch-based agro-industrial substrates. In this case, the glucosamine content was determined and converted to biomass (mg per g of dried substrate) using the equations deriving from the SSF on PDA. As depicted in Figure 4, the highest biomass production of 407.1 and 384.6 mg/g were detected in WG–PP and WG substrates for the AMRL 63 strain. The morel strain AMRL 52 was favored by the PP substrate, presenting its maximum biomass (215.5 mg/g) on the 20th day. These results are in agreement with those obtained during SmF and SSF using PDB and PDA media, respectively, which confirmed that the black morel AMRL 63 is able to produce higher biomass concentrations than the yellow morel AMRL 52. Papinutti and Lechner [58] reported that *M. esculenta* produced 54 mg/g biomass in wheat bran. Similar studies dealing with the evaluation of biomass production using other mushrooms species have been previously reported. Specifically, *L. edodes* reached up to 510.3 mg/g during SSF of bean stalks [4]. Among *Ganoderma* strains, *G. resinaceum* showed a maximum biomass production of 151.23 mg/g during SSF in spent mushroom substrate. Moreover, *Pleurotus ostreatus* and *Pleurotus pulmonarius* presented biomass concentrations up to 141.62 mg/g when cultivated on the same substrate [3]. It is worth noting that there is a positive correlation between growth rate and biomass production, as observed also in SSF using commercial substrates. These results are not in accordance with previous findings reporting that growth rate and biomass production of *Ganoderma* and *Pleurotus* strains were negatively related [3,4,59].

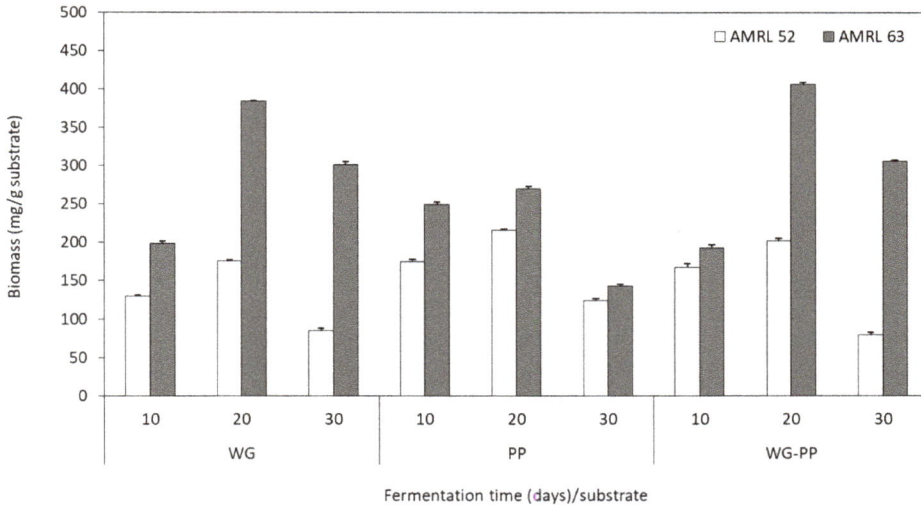

Figure 4. Estimated mycelium mass accumulation of yellow AMRL 52 and black AMRL 63 morel strains during solid-state fermentation on wheat grains (WG), potato peels (PP), and a mixture of them (WG–PP, 1:1).

The TP content of fermented agro-industrial substrates was similar for the black and yellow morels. Specifically, AMRL 52 and AMRL 63 mycelia achieved the highest TP content of 18.4% and 15.4% of dried biomass in PP and WG–PP, respectively. Previous studies have identified the bioactive compounds [60] deriving from *Morchella* mushrooms, showing that their functional properties are related to beneficial effects on human health [20,21,61]. The present results demonstrate the perspective for the production of bioactive compounds, such as glucosamine and polysaccharides, from *Morchella* sp. through the utilization of agro-industrial substrates.

4. Conclusions

The present study evaluated biomass production along with glucosamine and polysaccharides contents of black and yellow morel strains through the utilization of commercial and agro-industrial starch-based media. Glucosamine and polysaccharide mycelium contents were influenced by the age of the mycelia, presence of sclerotia, fermentation mode, type of substrate, and the strain. Linear regression equations between glucosamine and biomass of *Morchella* strains were reported for the first time. Biomass production and glucosamine content were found to be highly correlated, indicating that the determination of glucosamine content is a reliable indicator for the indirect estimation of *Morchella* biomass in SSF. In addition, high glucosamine and polysaccharide contents were correlated with high biomass production, presenting a R^2 value higher than 0.9 for the majority of fungal strains. Conclusively, *Morchella* strains were able to produce biomass rich in glucosamine and polysaccharides in SSF using starchy materials. These results suggest that the mycelium and sclerotia of *Morchella* sp. could be used as an alternative source of bioactive compounds. SSF have present some difficulties regarding the recovery of bioactive compounds [8]. Alternatively, fermented substrates (e.g., cereals, fruit pomace) enriched with bioactive compounds can be directly utilized as food supplements. This has already been suggested for fermented solids rich in polyunsaturated fatty acids [62], thus it could be expanded for other bioactive compounds, including chitin and polysaccharides using mycelium from edible fungi. In addition, it has been indicated that the supplementation of culture media with various nutrients, including vegetable oils, improved biomass production and its glucosamine

Foods **2019**, *8*, 239

content [8]. Hence, optimization of fermentation conditions and exploitation of other culture media could contribute to higher glucosamine and polysaccharide contents in fungal mycelium.

Author Contributions: Conceptualization, A.P. (Aikaterini Papadaki) and P.D.; methodology, A.P. (Aikaterini Papadaki), P.D., and A.P. (Antonios Philippoussis); writing—original draft preparation, A.P. (Aikaterini Papadaki) and A.P. (Antonios Philippoussis); writing—review and editing, A.P. (Aikaterini Papadaki), P.D., and S.P.; supervision, A.P. (Antonios Philippoussis) and P.D.

Funding: This research received no external funding.

Conflicts of Interest: The authors declare no conflict of interest.

References

1. Diamantopoulou, P.; Philippoussis, A. Cultivated Mushrooms: Preservation and Processing. In *Handbook of Vegetable Preservation and Processing*, 2nd ed.; Hui, Y.H., Evranuz, E.Ö., Bingöl, G., Erten, H., Jaramillo-Flores, M.E., Eds.; CRC Press: Boca Raton, FL, USA, 2015; pp. 495–525.

2. Carvajal, A.E.S.S.; Koehnlein, E.A.; Soares, A.A.; Eler, G.J.; Nakashima, A.T.A.; Bracht, A.; Peralta, R.M. Bioactives of fruiting bodies and submerged culture mycelia of *Agaricus brasiliensis* (*A. blazei*) and their antioxidant properties. *LWT-Food Sci. Technol.* **2012**, *46*, 493–499. [CrossRef]

3. Economou, C.N.; Diamantopoulou, P.A.; Philippoussis, A.N. Valorization of spent oyster mushroom substrate and laccase recovery through successive solid state cultivation of *Pleurotus*, *Ganoderma*, and *Lentinula* strains. *Appl. Microbiol. Biotechnol.* **2017**, *101*, 5213–5222. [CrossRef] [PubMed]

4. Philippoussis, A.; Diamantopoulou, P.; Papadopoulou, K.; Lakhtar, H.; Roussos, S.; Parissopoulos, G.; Papanikolaou, S. Biomass, laccase and endoglucanase production by *Lentinula edodes* during solid state fermentation of reed grass, bean stalks and wheat straw residues. *World J. Microbiol. Biotechnol.* **2011**, *27*, 285–297. [CrossRef]

5. Philippoussis, A.; Zervakis, G.; Diamantopoulou, P. Bioconversion of lignocellulosic wastes through the cultivation of the edible mushrooms *Agrocybe aegerita*, *Volvariella volvacea* and *Pleurotus* spp. *World J. Microbiol. Biotechnol.* **2001**, *17*, 191–200. [CrossRef]

6. Choong, Y.-K.; Ellan, K.; Chen, X.-D.; Mohamad, S.A. Extraction and Fractionation of Polysaccharides from a Selected Mushroom Species, *Ganoderma lucidum*: A Critical Review, Fractionation, Hassan Al- Haj Ibrahim. IntechOpen, 2018; pp. 39–60. Available online: https://www.intechopen.com/books/fractionation/extraction-and-fractionation-of-polysaccharides-from-a-selected-mushroom-species-ganoderma-lucidum-a (accessed on 2 June 2019). [CrossRef]

7. Hamed, I.; Özogul, F.; Regenstein, J.M. Industrial applications of crustacean by-products (chitin, chitosan, and chitooligosaccharides): A review. *Trend Food Sci. Technol.* **2016**, *48*, 40–50. [CrossRef]

8. Sitanggang, A.B.; Sophia, L.; Wu, H.S. Aspects of glucosamine production using microorganisms. *Int. Food Res. J.* **2012**, *19*, 393–404.

9. Ride, J.P.; Drysdale, R.B. A chemical method for estimating *Fusarium oxysporum* f. *lycopersici* in infected tomato plants. *Physiol. Plant Pathol.* **1971**, *1*, 409–420. [CrossRef]

10. Zdarta, J.; Meyer, A.S.; Jesionowski, T.; Pinelo, M. A General Overview of Support Materials for Enzyme Immobilization: Characteristics, Properties, Practical Utility. *Catalysts* **2018**, *8*, 92. [CrossRef]

11. Tan, C.; Feng, B.; Zhang, X.; Xia, W.; Xia, S. Biopolymer-coated liposomes by electrostatic adsorption of chitosan (chitosomes) as novel delivery systems for carotenoids. *Food Hydrocoll* **2016**, *52*, 774–784. [CrossRef]

12. Cheung, R.C.F.; Ng, T.B.; Wong, J.H.; Chan, W.Y. Chitosan: An Update on Potential Biomedical and Pharmaceutical Applications. *Mar. Drugs* **2015**, *13*, 5156–5186. [CrossRef]

13. A Worldwide Market with a Strong Demand. Available online: http://sflyproteins.com/a-worldwide-market-with-a-strong-demand/ (accessed on 2 June 2019).

14. Wu, T.; Zivanovic, S.; Draughon, F.A.; Sams, C.E. Chitin and Chitosan Value-Added Products from Mushroom Waste. *J. Agric. Food Chem.* **2004**, *52*, 7905–7910. [CrossRef] [PubMed]

15. Di Lena, G.; Annibate, A.D.; Sermanni, G.G. Influence of the age and growth conditions on the mycelial chitin content of *Lentinus edodes*. *J. Basic Microbiol.* **1994**, *34*, 11–16. [CrossRef]

16. Prasad, P.; Chauhan, K.; Kandari, L.S.; Maikhuri, R.K.; Purohit, A.; Bhatt, R.P.; Rao, K.S. *Morchella esculenta* (Guchhi): Need for scientific intervention for its cultivation in Central Himalaya. *Curr. Sci.* **2002**, *82*, 1098–1100.

17. Volk, T.J.; Leonard, T.J. Cytology of the life-cycle of *Morchella*. *Mycol. Res.* **1990**, *94*, 399–406. [CrossRef]

18. Winder, R.S. Cultural studies of *Morchella elata*. *Mycol. Res.* **2006**, *110*, 612–623. [CrossRef] [PubMed]

19. Güller, P.; Arkan, O. Cultural Characteristics of *Morchella esculenta* Mycelium on Some Nutrients. *Turk. J. Biol.* **2000**, *24*, 783–794.

20. Xu, N.; Lu, Y.; Hou, J.; Liu, C.; Sun, Y. A Polysaccharide Purified from *Morchella conica* Pers. Prevents Oxidative Stress Induced by H_2O_2 in Human Embryonic Kidney (HEK) 293T Cells. *Int. J. Mol. Sci.* **2018**, *19*, 4027. [CrossRef]

21. Yang, C.; Zhou, X.; Meng, Q.; Wang, M.; Zhang, Y.; Fu, S. Secondary Metabolites and Antiradical Activity of Liquid Fermentation of *Morchella* sp. Isolated from Southwest China. *Molecules* **2019**, *24*, 1706. [CrossRef]

22. Liu, C.; Sun, Y.; Mao, Q.; Guo, X.; Li, P.; Liu, Y.; Xu, N. Characteristics and Antitumor Activity of *Morchella esculenta* Polysaccharide Extracted by Pulsed Electric Field. *Int. J. Mol. Sci.* **2016**, *17*, 986. [CrossRef]

23. Lau, B.F.; Abdullah, N. Sclerotium-Forming Mushrooms as an Emerging Source of Medicinals: Current perspectives. In *Mushroom Biotechnology, Developments and Applications*; Petre, M., Ed.; Academic Press: Cambridge, MA, USA, 2016; pp. 111–136.

24. Gilbert, F. The submerged culture of *Morchella*. *Mycologia* **1960**, *52*, 201–209. [CrossRef]

25. Kaul, T.N. Physiological studies on *Morchella* spp. I. Carbon utilization. *Bull. Bot. Soc. Bengal.* **1978**, *31*, 35–42.

26. Kosaric, N.; Miyata, N. Growth of morel mushroom mycelium in cheese whey. *J. Dairy Res.* **1981**, *48*, 149–162. [CrossRef]

27. Philippoussis, A.; Balis, C. Studies on the morphogenesis of sclerotia and subterranean mycelial network of ascocarps in "*Morchella*" species. In *Science and Cultivation of Edible Fungi, Proceedings of the 14th international congress on the science and cultivation of edible fungi, Oxford, England*; Elliott, T.J., Ed.; A.A. Balkema: Rotterdam, the Netherlands, 1995; pp. 847–855.

28. Ang, T.N.; Ngoh, G.C.; Chua, A.S.M. Development of a novel inoculum preparation method for solid-state fermentation-Cellophane film culture (CFC) technique. *Ind. Crop. Prod.* **2013**, *43*, 774–777. [CrossRef]

29. Araujo, A.A.; Roussos, S. A technique for mycelial development of ectomycorrhizal fungi on agar media. *Appl. Biochem. Biotechnol.* **2002**, *98*, 311–318. [CrossRef]

30. Robson, G.D.; Bell, S.D.; Kuhn, P.J.; Trinci, A.P.J. Glucose and penicillin concentrations in agar medium below fungal colonies. *J. Gen. Microbiol.* **1987**, *133*, 361–367. [CrossRef] [PubMed]

31. Dantigny, P.; Guilmart, A.; Bensoussan, M. Basis of predictive mycology. *Int. J. Food Microbiol.* **2005**, *100*, 187–196. [CrossRef] [PubMed]

32. Baranyi, J.; Roberts, T.A.; McClure, P. A non-autonomous differential equation to model bacterial growth. *Food Microbiol.* **1993**, *10*, 43–59. [CrossRef]

33. Miller, G.L. Use of dinitrosalicylic acid reagent for determination of reducing sugars. *Anal. Chem.* **1959**, *31*, 426–428. [CrossRef]

34. Dubois, M.; Gilles, K.A.; Hamilton, J.K.; Rebers, P.A.; Smith, F. Colorimetric method for determination of sugars and related substances. *Anal. Chem.* **1956**, *28*, 350–356. [CrossRef]

35. Scotti, C.T.; Vergoignan, C.; Feron, G.; Durand, A. Glucosamine measurement as indirect method for biomass estimation of *Cunninghamella elegans* grown in solid state cultivation conditions. *Biochem. Eng. J.* **2001**, *7*, 1–5. [CrossRef]

36. Bailey, R.W. The reaction of pentoses with anthrone. *Biochem. J.* **1958**, *68*, 669–672. [CrossRef] [PubMed]

37. De Bruyn, J.W.; Van Keulen, H.A.; Ferguson, J.H.A. Rapid method for the simultaneous determination of glucose and fructose using anthrone reagent. *J. Sci. Food Agric.* **1968**, *19*, 597–601. [CrossRef]

38. Desgranges, C.; Vergoignan, C.; Georges, M.; Durand, A. Biomass estimation in solid state fermentation I. Manual biochemical methods. *Appl. Microbiol. Biotechnol* **1991**, *35*, 200–205. [CrossRef]

39. Roche, N.; Venague, A.; Desgranges, C.; Durand, A. Use of chitin measurement to estimate fungal biomass in solid state fermentation. *Biotechnol. Adv.* **1993**, *11*, 677–683. [CrossRef]

40. Brock, D.T. Studies on the Nutrition of *Morchella esculenta* Fries. *Mycologia* **1951**, *43*, 402–422. [CrossRef]

41. Stott, K.; Mohammed, C. *Specialty Mushroom Production Systems: Maitake and Morels. A report for the Rural Industries Research and Development Corporation*; Rural Industries Research and Development Corporation: Barton, Australia, 2004.

42. Zhang, G.P.; Zhang, F.; Ru, W.M.; Han, J.-R. Solid-state fermentation of cornmeal with the ascomycete *Morchella esculenta* for degrading starch and upgrading nutritional value. *World J. Microbiol. Biotechnol.* **2010**, *26*, 15. [CrossRef]

43. Xing, Z.; Sun, F.; Liu, J. Studies on the submerged-cultured conditions of *Morchella conica*. *Acta Edulis Fungi* **2004**, *11*, 38–43.

44. Buswell, A.J.; Chang, S. Biomass and extracellular hydrolytic enzyme production by six mushroom species grown on soybean waste. *Biotechnol. Lett.* **1994**, *16*, 1317–1322.

45. Bensoussan, M.; Tisserand, E.; Kabbaji, W.; Roussos, S. Partial characterization of aroma produced by submerged culture of morel mushroom mycelium. *Cryptog. Mycol.* **1995**, *16*, 65–75.

46. Meng, F.; Liu, X.; Jia, L.; Song, Z.; Deng, P.; Fan, K. Optimization for the production of exopolysaccharides from *Morchella esculenta* SO-02 in submerged culture and its antioxidant activities in vitro. *Carbohydr. Polym.* **2010**, *79*, 700–704. [CrossRef]

47. Xu, H.; Sun, L.-P.; Shi, Y.-Z.; Wu, Y.-H.; Zhang, B.; Zhao, D.-Q. Optimization of cultivation conditions for extracellular polysaccharide and mycelium biomass by *Morchella esculenta* As51620. *Biochem. Eng. J.* **2008**, *39*, 66–73. [CrossRef]

48. Sparringa, A.R.; Owens, D.J. Glucosamine content of tempe mould, *Rhizopus oligosporus*. *Int. J. Food Microbiol.* **1999**, *47*, 153–157. [CrossRef]

49. Ruíz-Herrera, J.; Osorio, E. Isolation and chemical analysis of the cell wall of *Morchella* sp. *Antonie van Leeuwenhoek* **1974**, *40*, 57–64. [CrossRef] [PubMed]

50. Broecker, F.; Seeberger, P.H. Identification and Design of Synthetic B Cell Epitopes for Carbohydrate-Based Vaccines. *Methods Enzymol.* **2017**, *597*, 311–334. [PubMed]

51. Tietel, Z.; Masaphy, S. True morels (*Morchella*)—Nutritional and phytochemical composition, health benefits and flavor: A review. *Crit. Rev. Food Sci. Nutr.* **2018**, *58*, 1888–1901. [CrossRef] [PubMed]

52. Zervakis, G.; Philippoussis, A.; Ioannidou, S.; Diamantopoulou, P. Mycelium growth kinetics and optimal temperature conditions for the cultivation of edible mushroom species on lignocellulosic substrates. *Folia Microbiol.* **2001**, *46*, 231. [CrossRef]

53. Masaphy, S. Effect of CaCO₃ on *Morchella* growth and sclerotia formation. *Int. Soc. Mushroom Sci.* **2004**, *16*. Available online: http://www.isms.biz/download/volume-16-part-1-article-14-effect-of-caco3-on-morchella-growth-and-sclerotia-formation/ (accessed on 2 June 2019).

54. Reeslev, M.; Kjøller, A. Comparison of biomass dry weights and radial growth rates of fungal colonies on media solidified with different gelling compounds. *Appl. Environ. Microbiol.* **1995**, *61*, 4236–4239.

55. Petre, V.; Petre, M.; Rusea, I.; Stănică, F. Biotechnological recycling of fruit tree wastes by solid-state cultivation of mushrooms. In *Mushroom Biotechnology, Developments and Applications*; Petre, M., Ed.; Academic Press: Cambridge, MA, USA, 2016; pp. 19–29.

56. Cheung, P.C.K.; Lee, M.Y. Fractionation and Characterization of Mushroom Dietary Fiber (Nonstarch Polysaccharides) as Potential Nutraceuticals from Sclerotia of *Pleurotus tuber-regium* (Fries) Singer. *J. Agric. Food Chem.* **2000**, *48*, 3148–3151. [CrossRef]

57. Alvarado-Castillo, G.; Mata, G.; Pérez-Vázquez, A.; Martínez-Carrera, D.; Tablada, M.E.N.; Gellardo-López, F.; Osorio-Acosta, F. *Morchella* sclerotia production through grain supplementation. *Interciencia* **2011**, *36*, 768–773.

58. Papinutti, L.; Lechner, B. Influence of the carbon source on the growth and lignocellulolytic enzyme production by *Morchella esculenta* strains. *J. Ind. Microbiol. Biotechnol.* **2008**, *35*, 1715–1721. [CrossRef] [PubMed]

59. Philippoussis, A.; Diamantopoulou, P. Exploitation of the biotechnological potential of agro-industrial by-products through mushroom cultivation. In *Mushroom Biotechnology and Bioengineering*; Petre, M., Berovic, M., Eds.; University of Pitesti: Bucharest, Romania, 2012; pp. 161–184.

60. Lo, Y.C.; Lin, S.Y.; Ulziijargal, E.; Chen, S.Y.; Chien, R.C.; Tzou, Y.J.; Mau, J.L. Comparative study of contents of several bioactive components in fruiting bodies and mycelia of culinary-medicinal mushrooms. *Int. J. Med. Mushrooms* **2012**, *14*, 357–363. [CrossRef] [PubMed]

61.   Pinto, M.R.; Barreto-Bergter, E.; Taborda, C.P. Glycoconjugates and polysaccharides of fungal cell wall and activation of immune system. *Braz. J. Microbiol.* **2008**, *39*, 195–208. [CrossRef] [PubMed]
62.   Ochsenreither, K.; Glück, C.; Stressler, T.; Fischer, L.; Syldatk, C. Production Strategies and Applications of Microbial Single Cell Oils. *Front. Microbiol.* **2016**, *7*, 1539. [CrossRef] [PubMed]

![foods logo] *foods*

MDPI

Article

Utilization of Clarified Butter Sediment Waste as a Feedstock for Cost-Effective Production of Biodiesel

Alok Patel [1,2,*], Km Sartaj [2], Parul A. Pruthi [2], Vikas Pruthi [2] and Leonidas Matsakas [1]

[1]   Biochemical Process Engineering, Division of Chemical Engineering, Department of Civil,
     Environmental and Natural Resources Engineering, Luleå University of Technology, 971 87 Luleå, Sweden
[2]   Molecular Microbiology Laboratory, Biotechnology Department, Indian Institute of Technology
     Roorkee (IIT-R), Roorkee 247667, India
*   Correspondence: alok.kumar.patel@ltu.se; Tel.: +46-72-451-8694

Received: 12 June 2019; Accepted: 28 June 2019; Published: 29 June 2019

Abstract: The rising demand and cost of fossil fuels (diesel and gasoline), together with the need for sustainable, alternative, and renewable energy sources have increased the interest for biomass-based fuels such as biodiesel. Among renewable sources of biofuels, biodiesel is particularly attractive as it can be used in conventional diesel engines without any modification. Oleaginous yeasts are excellent oil producers that can grow easily on various types of hydrophilic and hydrophobic waste streams that are used as feedstock for single cell oils and subsequently biodiesel production. In this study, cultivation of *Rhodosporidium kratochvilovae* on a hydrophobic waste (clarified butter sediment waste medium (CBM)) resulted in considerably high lipid accumulation (70.74% w/w). Maximum cell dry weight and total lipid production were 15.52 g/L and 10.98 g/L, respectively, following cultivation in CBM for 144 h. Neutral lipids were found to accumulate in the lipid bodies of cells, as visualized by BODIPY staining and fluorescence microscopy. Cells grown in CBM showed large and dispersed lipid droplets in the intracellular compartment. The fatty acid profile of biodiesel obtained after transesterification was analyzed by gas chromatography-mass spectrometry (GC–MS), while its quality was determined to comply with ASTM 6751 and EN 14214 international standards. Hence, clarified sediment waste can be exploited as a cost-effective renewable feedstock for biodiesel production.

Keywords: clarified butter sediment waste; hydrophobic substrates; oleaginous yeast; lipids; biodiesel; fatty acid methyl esters

1. Introduction

Current research seeks to find sustainable solutions to three eminent problems, namely global warming, energy crisis, and waste generation [1]. This implies increased attention towards biomass-based biofuels [2,3]. Several renewable transportation fuels already exist in the market; their aim being to diminish greenhouse gas emissions and reliance on petroleum-based fuels [4]. They include ethanol from corn and sugarcane, as well as biodiesel from vegetable crops, such as soy, palm, and rapeseed. However, the rising demand for transportation fuels, and the food versus fuel debate raises questions about their sustainability [5], sparking the need for alternative sources. Some non-edible feedstocks, such as waste cooking oils [6,7] and Soursop seed oil [8] are extensively used for biodiesel production but due to the presence of impurities, water, and high amount of free fatty acids, these substrates need extra pretreatments [7] or some modifications in transesterification reactions, such as a two-step process [8]. Most importantly, until conventional vehicle engines are replaced by electrical-powered ones, liquid biofuels remain the only viable renewable and sustainable fuel source for the transportation sector [9]. The vast amounts of liquid fuel required for transportation, create an important incentive towards increased production of renewable liquid fuels. Among the various alternatives, such as ethanol, biogas, and biodiesel, the latter holds an advantage as it can be directly used in conventional diesel engines without any modifications [5]. Chemically, biodiesel is a mixture of fatty acid methyl esters (FAMEs) and can be produced by using any type of fatty acid, regardless of its origin [10]. Oleaginous microorganisms, capable of accumulating high amounts of intracellular lipids, offer a sustainable solution to the use of edible oils for biodiesel production. Such microorganisms can synthesize lipids in the form of droplets [11] and some of them can accumulate more than 60% of lipids in their cellular compartments, under optimized cultivation conditions [12].

According to the Organization of Petroleum Exporting Countries, global oil demand will rise to 109.4 million barrels per day, of which diesel alone is expected to increase by 5.7 million barrels per day by 2040 [13]. As a consequence, the need to replace fossil fuels with renewable sources means that demand for biodiesel will follow a similar trend. To meet this challenge, a low-cost production process should be established. However, a prominent issue in biofuel production is cost, particularly the high cost of feedstock (60–75% of the total cost when using oleaginous microorganisms) [14]. To reduce production costs, novel strains capable of microbial lipids production using cheap waste streams, such as prickly-pear juice, industrial fats, monosodium glutamate wastewater, and crude glycerol derived from yellow grease, have been introduced [15–17].

Among the various types of microorganisms (microalgae, fungi, bacteria, and yeasts), the latter has been extensively studied due to their ability to grow on different types of waste substrates obtained from industrial or agricultural processes [3]. They utilize preferably hydrophilic substrates, and synthesize lipids in their cytoplasmic compartment via the de novo pathway under nutrient-limited conditions [18]. In contrast, only yeasts capable of synthesizing lipids via the ex novo pathway can utilize hydrophobic substrates. *Yarrowia lipolytica* is one of the most studied oleaginous yeasts for ex novo lipid accumulation [19], together with *Cryptococcus*, *Rhodosporidium*, *Geotrichum*, and *Trichosporon*, which can all synthesize lipids from hydrophobic substrates. Several multigene families have been identified in these oleaginous yeasts that take part in the degradation of hydrophobic substrates at the cell surface and their assimilation into triacylglycerol (TAG) in the cellular compartment [20]. Uptake of hydrophobic substrates is started by extracellular lipases, which degrade such substrates into free fatty acids (FFA), prior to their internalization by active transport or gradient-based diffusion [20,21]. To facilitate FFA internalization or increase contact between the cell's surface and the substrate, oleaginous yeasts' surfaces are modified with protrusions [22,23]. After internalization, FFA are either consumed for cell growth or are taken up by the ex novo lipid synthesis pathway. In the latter case, they are reserved in the form of lipid bodies with same compositional contents or are enzymatically modified to yield fatty acids present in the medium [24]. β-oxidation plays a crucial role after assimilation of FFAs to cover the energy demands for cell growth and repair or synthesis of other metabolites [23].

The main hydrophobic substrates used for microbial cultivation are FFA from various sources, such as wastewater from food industries and dairy industries, waste cooking oil, and waste fish oils [20,21,25,26]. Millions of tons of waste cooking oil and animal fats are produced per year; these are responsible for environmental pollution when directly disposed into landfill sites or water bodies. Moreover, there is still no efficient method for the processing of used oils from households [27]. Therefore, utilization of hydrophobic substrates by oleaginous yeasts offers a cost-effective approach for biodiesel production, as well as a solution for managing such detrimental waste [27,28].

Ghee or clarified butter is an important global meal ingredient [29]. Clarified butter sediment waste or ghee residue is one of the largest byproducts of the dairy industry, with almost 91,000 tons per annum being produced in India alone [30]. It is generated after coagulation of the solid, non-fat part of the cream of milk, during the preparation of ghee [31]. It can be easily separated from clarified butter by filters, muslin cloth or continuous centrifugal clarifiers. The yield of clarified butter sediment waste depends on the raw materials and the method of preparation [32]. The maximum reported average yield is 12% of the initial raw material through the direct creamery method, followed by 3.7% yield in creamery butter and desi butter methods [33]. The chemical composition of clarified butter sediments waste varies according to the initial raw cream quality and method of preparation; it usually consists of fat (32–70%), protein (12–39%), moisture (8–30%), lactose (2–14%), and ash (1–8%) [29]. The residue also contains a high quantity of phenolic compounds that might be harmful for human consumption [34]. As it is usually discarded as waste into the environment, butter sediments waste could represent a low-cost waste feedstock for biodiesel production and, as such, could be less of a burden for the environment [35]. Previously, ghee residue/clarified butter sediment waste was utilized for the cost-effective production of microbial lipases through solid-state fermentation. For example, *Bacillus subtilis* (NCIM 2063) and *Proteus* spp. produced high amounts of lipases, corresponding to 35.93 U/mg and 41.27 U/mg, respectively, when cultivated on solid ghee residue [31]. Clarified butter sediment waste from the dairy industry has also been used for cultivating bacteria (*Bacillus sphaericus* and *Bacillus thuringiensis* serovar *israelensis*) with biomass yields of 9.7 g/L and at a lower cost than with a conventional nutrient yeast extract salt medium [35].

In the present study, the oleaginous yeast *Rhodosporidium kratochvilovae* HIMPA1 was used to assess the feasibility of utilizing clarified butter sediment waste as a low-cost hydrophobic substrate, particularly in comparison with conventional glucose synthetic medium. The current work aims to deliver a sustainable and environment-friendly process for the valorization of clarified butter sediment waste medium (CBM) in the production of microbial lipids that are suitable for use as biodiesel (Figure 1).

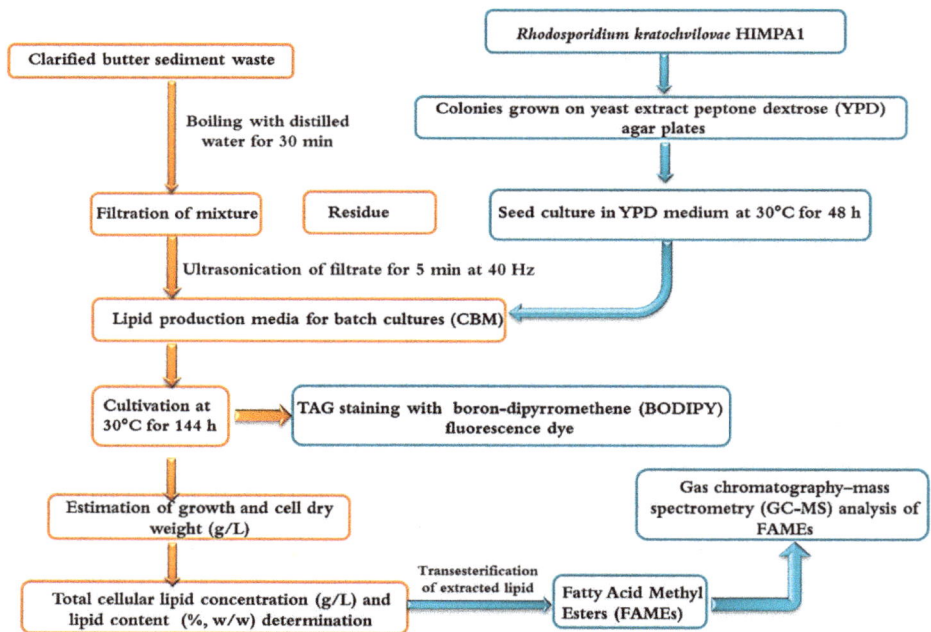

Figure 1. Schematic representation for the utilization of clarified butter sediment waste as a feedstock to cultivate *R. kratochvilovae* HIMPA1 for biodiesel production.

2. Materials and Methods

2.1. Materials

The chemicals and reagents used in the present study were of analytical grade and procured from Merck, Mumbai, India. Media components for yeast cultivation were purchased from Hi-Media (Mumbai, India). BODIPY$_{505/515}$ (4,4-difluoro-1,3,5,7-tetramethyl-4-bora-3a, 4a-diaza-s-indacene) used to visualize the intracellular lipid droplets was procured from Invitrogen (Life Technology, Carlsbad, CA, USA).

2.2. Yeast Strain and Preparation of Seed Culture

R. kratochvilovae HIMPA1 (GenBank Acc. No. KF772881) used in the present study was previously isolated by our group from permafrost soils in the Himalayan region [36]. Seed culture was prepared by inoculating a loopful of colonies from yeast extract peptone dextrose (YPD) agar plates into YPD broth (100 mL) and incubating them at 30 °C for 48 h.

2.3. Preparation of CBM and Yeast Cultivation

A charred (brunt) light-to-dark-brown sediment was obtained during the clarification of the collected milk cream or butter. The initial substrate of clarified butter sediment waste was milk cream. Cream (500 g) was collected from the milk after concentrating the milk fat. The cream was then heated to the boiling temperature (110–120 °C) with continuous stirring for 2–3 h, until the clear liquid could be separated from the small dark brown particles. The clarified butter was separated from solid residues by filtration through a sieve. To determine the total fat present in the solid residue, 5 g was extracted with 10 mL of *n*-hexane, in triplicates, the solvent was evaporated, and the total fat was estimated gravimetrically. The protein content was analyzed by the Kjeldahl method. To utilize the clarified butter sediment waste as a substrate for yeast cultivation, 10 g of it was boiled in 100 mL distilled water for 30 min. The mixture was filtered using a mesh strainer, and the filtrate was ultrasonicated at 40 Hz for 5 min. After cooling the filtrate, appropriate amounts of Yeast Nitrogen Base (YNB) without ammonium sulfate (0.79 g/L) and the complete supplement mixture (CSM, 1.7 g/L) were added to the mixture, whose final volume was then adjusted to 98 mL by adding distilled water. The pH of the medium was adjusted to 6.8 before autoclaving at 121 °C for 15 min. The medium was then inoculated with 2% (*v/v*) of a seed culture previously grown for 48 h, and incubated at 30 °C with 180 rpm for 144 h. Residual fat during the cultivation period was monitored by repeated *n*-hexane extraction of the supernatant. *n*-hexane was then transferred to pre-weighed glass vials and kept in the oven until solvent evaporation was complete. Finally, the residual fat (g/L) was gravimetrically determined.

2.4. Determination of Cell Dry Weight and Lipid Content

For cell dry weight determination, 10 mL of the culture was harvested every 24 h and centrifuged at 5,000 rpm. The obtained pellets were washed with distilled water, followed by washing with *n*-hexane to remove the medium components, as well as the fat present in CBM. The pellets were dried overnight on a pre-weighed aluminium boat, in an oven set at 80 °C, and the cell dry weight was gravimetrically determined. Intracellular lipids were extracted by a modified Bligh and Dyer protocol previously described by our group [36]. Total lipid concentration was estimated gravimetrically in g/L, whereas lipid content (%, w/w) was estimated using the following equation (Y) = Total lipid concentration (g/L)/cell dry weight (g/L) × 100 [36].

2.5. Determination of TAG Synthesis in R. kratochvilovae HIMPA1 Cells

Morphology and TAG synthesis ability of *R. kratochvilovae* HIMPA1 cultivated in CBM were observed by live fluorescence microscopy. Samples of 1 mL were harvested and washed with distilled water to remove the residual medium. The samples were washed again with *n*-hexane to remove the extra unutilized fats attached to the cells' outer surface, which could have interfered with fluorescence microscopy. The pellets were resuspended in 50 μL distilled water and 2 μL of BODIPY$_{505/515}$ (0.1 mg/mL in DMSO) was added to the solution in amber Eppendorf tubes to avoid photobleaching of the dye. The tubes were incubated for 5 min at room temperature. Bright field and fluorescence microscopy images were taken with an inverted fluorescence microscope (EVOS-FL, Thermo Fisher Scientific, Waltham, MA, USA) equipped with an EVOS LED light cube GFP filter (470/22 nm Excitation; 510/42 nm Emission) [37] at a 60× magnification.

2.6. Estimation of the Fatty Acid Profile

The extracted lipids were transesterified with acid catalysts (0.6 M HCl in methanol), as previously described [38]. The reaction was carried out in polytetrafluoroethylene (PTFE)-capped glass vials at 85 °C for 1 h and FAME were extracted with *n*-hexane after cooling the reaction mixture to the room temperature. The transesterified products were analyzed by gas chromatography-mass spectrometry (GC–MS) (Agilent, Santa Clara, CA, USA) using a DB-5MS capillary column (30 m × 0.25 mm ID and 0.25 μm film thickness). The column was initially heated at 50 °C for 1.5 min and the temperature was ramped at 25 °C/min to 180 °C, where it was maintained for 1 min. Then, the temperature was further increased to 280 °C, at a rate of 10 °C/min and maintained there for 1 min. Finally, the temperature was ramped to 250 °C at a rate of 15 °C/min and maintained at this temperature for 3 min. A total of 1 μL of the sample was loaded by autosampler in the splitless mode at 250 °C. The peaks were examined with the help of electron ionization (70 eV; scan mode 50–600 *m/z*). The temperature of the ion source was set up at 200 °C, while the mass transfer line was maintained at 250 °C. The identification of peaks was carried out by a Probability-Based Matching (PBM) library search in the MS spectral database.

2.7. Theoretical Analysis of the Biodiesel's Properties

The biodiesel's properties were estimated by the following empirical formulae [39]:
Saponification value (SV; mg KOH) = $\sum$ 560 (% FC)/M
Iodine value (IV; gI$_2$/100 g) = $\sum$ 254 DB × % FC/M
Cetane number (CN) = 46.3 + 5.458/SV − (0.255 × IV)
Degree of unsaturation (DU; % weight) = MUFA + (2 × PUFA)
Long chain saturation factor (LCSF; % wt.) = (0.1 × C16) + (0.5 × C18) + (1 × C20) + (1.5 × C22) + (2 × C24)
Cold filter plugging point ((CFPP); °C) = (3.417 × LCSF) − 16.477
Higher heating value (HHV; MJ/Kg) = 49.43 − 0.041 (SV) − 0.015 (IV)
Kinematic viscosity (ln KV at 40 °C in mm^2/s) = −12.503 + 2.496 × ln ($\sum$M) − 0.178 × $\sum$DB
Density = 0.8463 + 4.9/$\sum$M + 0.0118 × $\sum$DB
Oxidative stability (OS; h) = 117.9295/(wt. % C18:2 + wt. % C18:3) + 2.5905
where M = molecular mass of each fatty acid component, DB = number of double bonds, FC = fatty acid component (%), MUFA = monounsaturated fatty acids (%), PUFA = polyunsaturated fatty acids (%).

2.8. Statistical Analysis

All experiments were performed in triplicates and the results were presented as the standard deviation of three independently recorded values.

3. Results and Discussion

3.1. Batch Cultivation of the Oleaginous Yeast in CBM

To reduce the cost accompanying the feedstocks for biodiesel production by oleaginous microorganisms, it is important to identify strains capable of growing on low-cost substrates [3]. While oleaginous yeasts generally prefer to utilize hydrophilic substrates such as sugars and synthesize lipids via the de novo synthesis pathway [11], some can also efficiently metabolize hydrophobic substrates via ex novo lipid synthesis [20]. In the de novo pathway, the carbon substrate is converted into the lipid precursors acetyl- and malonyl-CoA by the TCA cycle, and finally forms C14–C18 long fatty acid chains, depending on the type of microorganisms. In contrast, hydrophobic substrates are first degraded at the surface of cells in the ex novo pathway and are incorporated as such, or with modifications [40].

In this study, CBM was used as a hydrophobic waste substrate to cultivate *R. kratochvilovae*. A total of 500 g of milk cream was used to prepare 466 g of clarified butter or ghee (93.2% of total milk cream) and 34 g of solid residue (6.8% of total milk cream) (Figure 2). The w/w composition of the solid residue was 39.1% fat, 19.2% protein, 30.1% moisture, 2.5% lactose, and 7.3% ash. The solid residue (10 g) was boiled in 100 mL of distilled water and sonicated to facilitate fat dispersion in the medium, thus yielding 2.84% w/v of fat and 7.8 mg/mL of protein.

Solid sediment waste

Clarified butter/Ghee

Milk cream

Figure 2. Production of clarified butter from milk cream; a charred (brunt) light-to-dark-brown solid sediment was obtained during the clarification process.

Hydrophobic substrates usually form unstable emulsions in aqueous solutions or float on the surface, as shown in Figure 3. To tackle this issue, oleaginous yeasts often secrete extracellular emulsifiers, such as liposan, or extracellular lipases that facilitate substrate degradation [41]. Nevertheless, to enhance the contact area between hydrophobic substrates and the cell surface, some external emulsifiers are often used to reduce the size of the hydrophobic substrates [42]. For example, in one study, oleaginous yeast *Trichosporon dermatis* CH007 was cultivated on a soybean oil-based medium along with Tween 80, Span 80, Tween 60, and OP-10 as external emulsifiers [43]. The addition of emulsifiers had a minor negative effect on the final lipid production. More importantly, though, it interfered with the quantification of residual hydrophobic substrates in the medium, due to its solubility in the organic solvent that was used for recovery of the residual hydrophobic substrates [43]. Here, CBM was ultrasonicated to reduce the fat size and make the hydrophobic substrates available in the form of easily accessible tiny droplets (Figure 3). Ultrasonication has been previously applied to facilitate the growth of *Cryptococcus curvatus* on waste cooking oil, whereby lipid accumulation was found to increase from 12.21 ± 1.34 g/L (unsonicated) to 20.34 g/L [44].

Unsonicated   Sonicated

0 h   72 h   144 h

Figure 3. Appearance of fat from ultrasonicated clarified butter sediment waste medium (CBM) during utilization by *R. kratochvilovae* over a cultivation period of 144 h.

Time-course results pertaining to the cell dry weight, total lipid concentration, and lipid content for *R. kratochvilovae* cultivated in CBM are presented in Figure 4. No lag phase was observed during the initial stage of cultivation, and the cell dry weight and lipid accumulation was found to increase linearly, over a period of 144 h. The highest cell dry weight, total lipid concentration, and lipid content were 15.52 g/L, 10.98 g/L, and 70.74% (w/w), respectively after 144 h (Figure 4). Lipid content increased from 37.93% (w/w) at 24 h to 54.26% (w/w) after 72 h of cultivation. Lipid-free biomass was almost constant between 72 h and 144 h of cultivation, possibly due to simultaneous degradation and intake of fat at the cell surface. Indeed, some oleaginous yeasts, such as *Y. lipolytica*, has been shown to possess extracellular lipolytic and proteolytic activities when grown on hydrophobic substrates [20,45]. Moreover, even the filamentous fungus *Aspergillus* sp. ATHUM 3482 cultivated on a carbon-limited hydrophobic medium with waste cooking olive oil was reported to synthesize 64% (w/w) of lipids in its cellular compartment [46]. Additionally, cultivation of the oleaginous yeast *Rhodotorula glutinis* TISTR 5159 on palm oil mill effluent—another hydrophobic material—yielded 6.32 g/L of biomass, along with a 32.63% lipid content [47]. Other oleaginous yeasts and the corresponding hydrophobic waste materials that were cultivated are listed in Table 1.

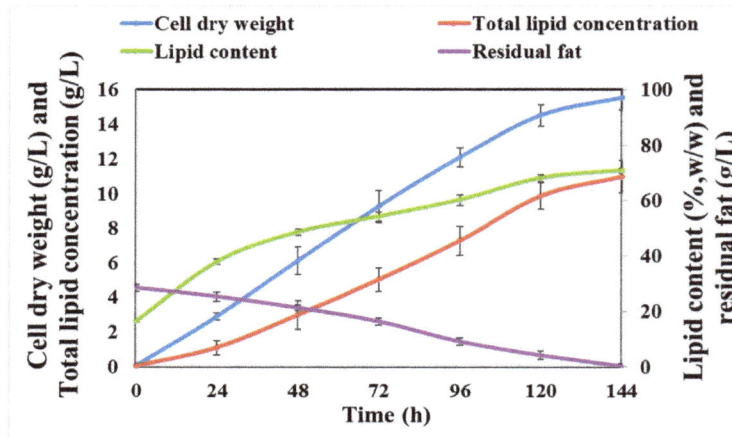

Figure 4. Time course showing changes in the cell dry weight, lipid content, lipid concentration, and fat consumption by *R. kratochvilovae* cultivated in CBM over a period of 144 h.

Oleaginous yeasts can resort to de novo lipid synthesis when grown in the presence of hydrophilic carbon and under nitrogen-limited conditions [48]. Instead, ex novo lipid synthesis is entirely independent of nitrogen concentration when hydrophobic materials are supplied as the cultivating medium [49]. Here, cell dry weight and lipid accumIMPulation following growth of *R. kratochvilovae* HIMPA1 on the hydrophobic substrates (CBM) were considerably higher, compared to those obtained on hydrophilic substrates such as a glucose synthetic medium (GSM), containing 70 g/L of glucose and 5 g/L of ammonium sulfate [18]. The maximum cell dry weight and lipid concentration on the GSM were 13.26 g/L and 6.2 g/L, respectively, corresponding to only a 46.76% (*w/w*) lipid content [18]. During de novo lipid accumulation, oleaginous yeasts prefer to utilize glucose due to its high-energy content (~2.8 kJ/mol) [50]. However, de novo lipid accumulation is also affected by the amount of carbon source, the presence of other medium components (nitrogen, phosphorus, and sulfur), and their ratio to the carbon source (C/N, C/P, and C/S) [11,51,52]. In our previous study, the concentration of nitrogen and phosphorus was optimized so that when this oleaginous yeast was cultivated in 7% glucose with 0.1 g/L of ammonium sulfate, the highest cell dry weight and lipid concentrations were 9.23 g/L and 5.51 g/L, respectively, corresponding to 59.69% (w/cell dry weight) lipid content [18]. A similar growth pattern was observed when *Rhodosporidium kratochvilovae* SY89 was cultivated in a glucose-based medium with cane molasses as a carbon source [53]. Both values were, however, lower than the 70.74% lipid content obtained here, after 144 h of cultivation on CBM. During de novo lipid accumulation, isocitrate dehydrogenase (ICDH) is not accessible to convert isocitrate to α-ketoglutarate as ICDH is totally dependent on the concentration of AMP, and all AMP converts into inosine monophosphate, IMP by AMP-deaminase, during the condition of nitrogen starvation in the mitochondria [11]. The isocitrate accumulating in the mitochondria equilibrates with citrate via the citrate–malate–translocase system [54]. ATP–citrate lyase converts citrate to acetyl-Co-A and oxaloacetate. Finally, acetyl-Co-A and malonyl-Co-A act as lipid precursors for fatty acid chains between C14 and C16 [55]. In comparison, under similar nitrogen-limited conditions, conventional or non-oleaginous yeasts such as *Saccharomyces cerevisiae*, convert the carbon source into mannans and glucans [56]. However, the mechanisms of hydrophobic substrates assimilation by any oleaginous microorganisms are totally different from the hydrophilic substrate utilization. The assimilation of hydrophobic substrates usually starts with two different mechanisms, either by the surface-mediated transport or by direct interfacial transport [45]. After assimilation of these substrates, various pathways such as monoterminal alkane oxidation, β-oxidation, citrate, and glyoxylate cycle, work to degrade the hydrophobic substrates [21,57]. Although all of these pathways work in different compartments, such as ER, mitochondria, and peroxisome, the final destination is always β-oxidation in peroxisome [58]. The excess amount of hydrophobic substrates could be reserved in the form of lipid droplets [21].

Table 1. Oleaginous yeasts cultivated on the hydrophobic substrates.

Oleaginous Yeast	Hydrophobic Substrate	Cell Dry Weight (g/L)	Lipid Concentration (g/L)	Lipid Content (% w/w)	References
Trichosporon dermatis	Soybean oil (20 g/L)	26.7	11.6	43.4	[43]
Cryptococcus. curvatus	Sonicated waste cooking oil (20 g/L)	18.62	13.06	70.13	[44]
Rhodotorula glutinis	Waste cooking oils + crude glycerol	25.5		46 ± 5%	[59]
Yarrowia lipolytica	Industrial saturated fats (stearin)	12.5	6.8	54	[60]
Candida lipolytica 1094	Corn oil (18 g/L)		10.1	55	[61]
Yarrowia lipolytica mutant strains (YlB6, YlC7, and YlE1)	Waste cooking oil (100 g/L)	10.86, 7.1 and 5.84	5.97, 4.28, and 3.91	55, 60, and 67	[62]
Rhodosporidium kratochvilovae HIMPA1	CBM	15.52	10.98	70.74	This Study

CBM, clarified butter sediment waste medium.

3.2. Detection of TAG by Fluorescence Microscopy in R. kratochvilovae Cultivated in CBM

Synthesis of lipid droplets (also known as lipid bodies or adiposomes), their dynamics, size, number, and composition are totally dependent on the type of oleaginous microorganisms. Moreover, for the same strain, cultivation conditions including supplied nutrients affect the structure of lipid droplets [63]. Figure 5 shows the bright field and fluorescence images of lipid droplets in *R. kratochvilovae* HIMPA1 grown on CBM for 144 h. The image shows a direct correlation between the lipid droplets' size and TAG accumulation under ex novo lipid synthesis. The cell and lipid droplet size of this oleaginous yeast were shown to vary with changes in the concentration of nitrogen and phosphorus in GSM [18]. Specifically, when cells were grown under nitrogen-limited conditions, lipid droplets were perfectly rounded; whereas under phosphorus-limited conditions, they were more diffuse and irregular, likely due to impaired phospholipid synthesis [18] and, thus, an aberrant monolayer surrounding the droplets was observed [57,64]. However, these variations might also reflect the different mechanisms responsible for lipid droplet synthesis. These include the 'lensing model' and the 'bicelle formation model', which describe the genesis of lipid droplets in cellular compartments [57,64]. In the lensing model, lipid droplets covered by a monolayer, originated from a single cytosolic leaflet of the endoplasmic reticulum (ER) membrane; whereas, both leaflets of the ER membrane took part in the formation of the monolayer in the bicelle model, [65]. Hence, in the case of GSM-cultivated cells, the lipid droplets were uniform and rounded; whereas, in the case of CBM-grown cells, they appeared to be more dispersed (Figure 5).

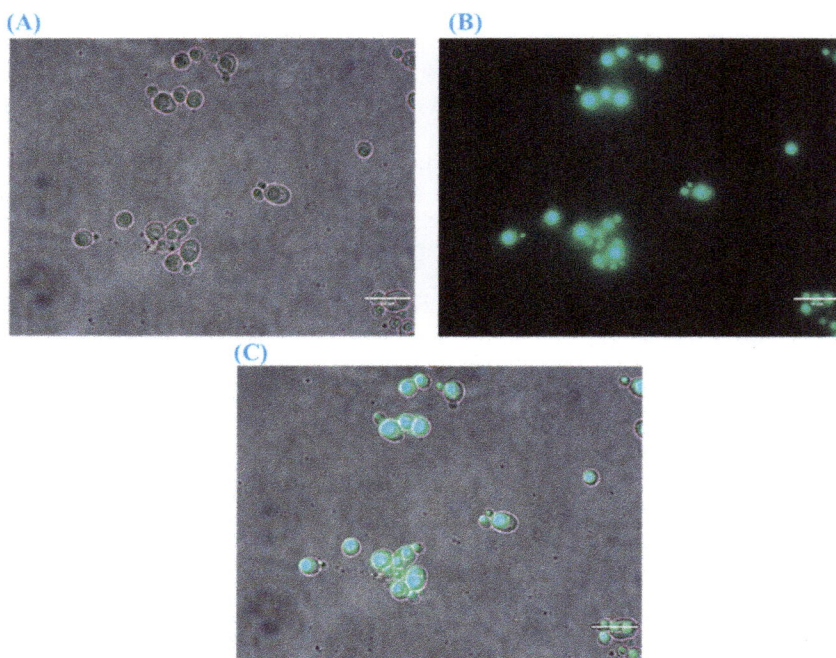

Figure 5. Bright field (**A**) and fluorescence (**B**) and merged images (**C**) showing lipid droplets accumulation in the oleaginous yeast *R. kratochvilovae* cultivated in CBM.

3.3. Fatty Acid Profile of R. kratochvilovae Cultivated in CBM

The lipids extracted from *R. kratochvilovae* cultivated in CBM were analyzed by GC–MS after transesterification (Figure 6). *R. kratochvilovae* HIMPA1 mainly synthesizes myristic acid (C14:0), palmitic acid (C16:0), stearic acid (C18:0), oleic acid (C18:1), along with linoleic acid (C18:2) and traces of linolenic acid (C18:3) [18]. The exact content of these fatty acids varies with the nature of the feedstock. For example, when this yeast was grown in pulp and paper industry effluent, the fatty acid content included C16:0 (21.86%), C18:0 (0.5%), C18:1 (45.43%), C18:2 (15.91%), and traces of arachidic acid (C20:0) (0.12%) [66]. In contrast, *R. kratochvilovae* HIMPA1 cultivated on hemp seed aqueous extract containing high amounts of lipids, resulted in 62.5% of saturated fatty acids (SFA), 37.5% of MUFAs, and the unusual carboceric fatty acid (C27:0) [36]. Here, a completely different profile was detected when this oleaginous yeast was cultivated in CBM; this included a high degree of unsaturation (83.09%) that comprised mainly C16:0 (16.67%), C18:0 (15.37%), C18:1 (49.70%), C18:2 (14.35%), C18:3 (2.13%), C20:0 (0.56%), and eicosenoic acid (C20:1) (0.23%) (Figure 6). In comparison, GSM-grown cultures showed a high amount of saturated fats (60.35%), including C16:0 (34.79%), C18:0 (21.32%), and C14:0 (4.24%); and fewer unsaturated ones, such as C18:1 (23.15%) and C18:2 (2.23%) [18].

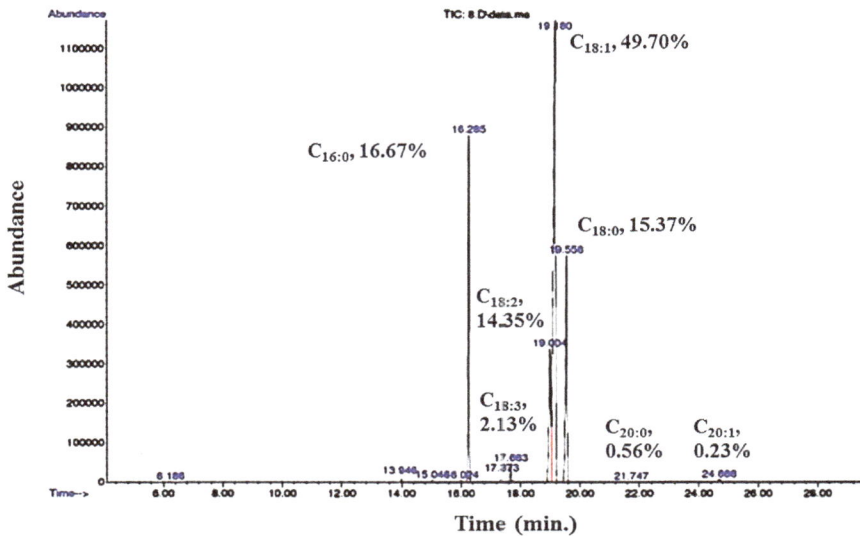

Figure 6. GC-MS chromatogram showing the fatty acid profile of *R. kratochvilovae* cultivated in CBM.

Biodiesel quality was affected by several factors, such as fatty acids composition of the feedstock (e.g., chain length and number, and position and isomers of double-bonds), production process, refining process, and post-production parameters [67–69]. The properties of biodiesel obtained from the CBM- and GSM-cultivated cells are listed in Table 2, where they are also compared with the two most important international standards—the American Society of Testing and Materials ASTM 6751, and the European Standard EN 14214.

Table 2. Theoretical estimation of biodiesel properties from the CBM-cultivated cells.

Biodiesel Properties	CBM	GSM [18]	Standard Biodiesel Parameters	
			ASTM D6751-02	EN 14214
Degree of unsaturation	83.09	27.61	-	-
Saponification value (mg/g)	199.78	160.26	0.50 min	0.50 min
Iodine value (mgI$_2$/100 g)	76.97	23.71	-	120 (max)
Cetane number	56.30	74.30	47 min	51 min
Long chain saturated factor	9.91	14.13	-	-
Cold filter plugging point	14.66	31.83	-	-
Cloud point (°C)	3.77	13.30	-	-
Pour point (°C)	−2.72	7.625	-	-
Oxidation stability (h)	9.70	55.47	3 min	6 min
Higher heating value (MJ/kg)	40.08	42.50	-	-
Kinematic viscosity (mm^2/s)	3.96	4.52	1.9–6.0	3.5–5
Density (g/cm^3)	0.87	0.87	-	0.86–0.90

GSM, glucose synthetic medium.

A key parameter affecting biodiesel performance in cold environments is the CFPP, which depends strongly on the degree of unsaturation. CBM-grown cells contained a high quantity of MUFAs (48.90%), resulting in a lower CFPP (14.66 °C) than FAME obtained from the GSM-cultivated cells (31.83 °C). Oxidative stability is an essential fuel property influenced by the presence of large amounts of SFA in oil feedstock. Increased number of double-bonds or unsaturation in the fatty acids led to a higher degree of autoxidation of biodiesel and, consequently, a reduced shelf life [70]. Here, biodiesel obtained from GSM-grown cells was more stable (55.47 h) than that obtained from cells grown in CBM (9.70 h); however, both values were within the accepted international standards. In general, biodiesel has a 12% lower HHV than diesel (39.57–41.33 MJ/kg) [71]. This led to a higher consumption of biodiesel to attain a similar yield as that of conventional diesel [71,72]. Here, the biodiesel obtained from the CBM-grown cells had a slightly lower HHV (40.08 MJ/kg) than that obtained from the GSM-grown cells (42.50 MJ/kg). Kinematic viscosity determined the properties of fuel injection into the engine, with a highly viscous fuel having a larger droplet size and inferior vaporization. As shown in Table 2, both biodiesel types were well within the set international standards. In general, unsaturation in fatty acids enhanced the operability of biodiesel in a cold environment, due to a low CFPP; whereas their saturation prevented auto-oxidation of the fuel and improved the shelf life of biodiesel [73]. It is quite difficult to find a feedstock that meets all criteria for biodiesel; therefore, blending different biodiesels to achieve the optimum ratio of saturated to unsaturated fatty acids could be an effective solution [74,75].

4. Conclusions

Clarified sediment waste is one of the largest waste materials of the dairy industry. It is usually discarded in the surrounding environment or drained in water bodies at industrial, as well as domestic level, thus severely affecting the aquatic ecosystem and generating a putrid odor. Due to the presence of a high amount of fat, it cannot be degraded except by oleaginous microorganisms, which secrete extracellular lipases and engage in ex novo lipid synthesis. In this study, clarified sediment waste was successfully utilized as hydrophobic substrate by an oleaginous yeast, *R. kratochvilovae* HIMPA1, which synthesized 70.74% *w/w* lipids in its cellular compartment over a cultivation period of 144 h. The highest cell dry weight and total lipid concentration were 15.52 g/L and 10.98 g/L, respectively. The biodiesel produced after transesterification of the lipids extracted from these cultures satisfied international standards for biodiesel fuel. Hence, utilization of clarified sediment waste not only solves the problem of waste generation, but can be also exploited as a renewable feedstock for cost-effective biodiesel production.

Author Contributions: Conceptualization, A.P.; Data curation, V.P.; Funding acquisition, P.A.P.; Investigation, A.P.; Methodology, A.P. and K.S.; Project administration, P.A.P.; Resources, P.A.P.; Software, L.M.; Supervision, P.A.P. and V.P.; Visualization, L.M.; Writing—original draft, A.P. and K.S.; Writing—review & editing, P.A.P., V.P., and L.M.

Funding: This research was funded by the DBT, Government of India, (Sanction No. 102/IFD/SAN/3539/2011-2012; Grant No.: DBT-608-BIO) and SRF fellowship to A.P. from University Grants Commission, India (Grant No.: 6405-35-044).

Acknowledgments: We thank Supriya Patel for providing the clarified butter sediment waste that was used in this study.

Conflicts of Interest: The authors declare no conflict of interest.

References

1. Amorim, W.S.; Valduga, I.B.; Ribeiro, J.M.P.; Williamson, V.G.; Krauser, G.E.; Magtoto, M.K.; de Andrade Guerra, J.B.S.O. The nexus between water, energy, and food in the context of the global risks: An analysis of the interactions between food, water, and energy security. *Environ. Impact Assess. Rev.* **2018**, *72*, 1–11. [CrossRef]
2. Lin, L.; Gettelman, A.; Fu, Q.; Xu, Y. Simulated differences in 21st century aridity due to different scenarios of greenhouse gases and aerosols. *Clim. Chang.* **2018**, *146*, 407–422. [CrossRef]
3. Patel, A.; Arora, N.; Sartaj, K.; Pruthi, V.; Pruthi, P.A. Sustainable biodiesel production from oleaginous yeasts utilizing hydrolysates of various non-edible lignocellulosic biomasses. *Renew. Sustain. Energy Rev.* **2016**, *62*, 836–855. [CrossRef]
4. Jenniches, S. Assessing the regional economic impacts of renewable energy sources—A literature review. *Renew. Sustain. Energy Rev.* **2018**, *93*, 35–51. [CrossRef]
5. Saidur, R.; Boroumandjazi, G.; Mekhilef, S.; Mohammed, H.A. A review on exergy analysis of biomass based fuels. *Renew. Sustain. Energy Rev.* **2012**, *16*, 1217–1222. [CrossRef]

6. Bobadilla, M.C.; Fernández Martínez, R.; Lorza, R.L.; Gómez, F.S.; González, E.P.V. Optimizing biodiesel production fromwaste cooking oil using genetic algorithm-based support vector machines. *Energies* **2018**, *11*, 2995. [CrossRef]

7. Bobadilla, M.C.; Lorza, R.L.; Escribano-Garcia, R.; Gómez, F.S.; González, E.P.V. An improvement in biodiesel production from waste cooking oil by applying thought multi-response surface methodology using desirability functions. *Energies* **2017**, *10*, 130. [CrossRef]

8. Su, C.H.; Nguyen, H.C.; Pham, U.K.; Nguyen, M.L.; Juan, H.Y. Biodiesel Production from a Novel Nonedible Feedstock, Soursop (*Annona muricata* L.) Seed Oil. *Energies* **2018**, *11*, 2562. [CrossRef]

9. Daraei, M.; Thorin, E.; Avelin, A.; Dotzauer, E. Potential biofuel production in a fossil fuel free transportation system: A scenario for the County of Västmanland in Sweden. *Energy Procedia* **2019**, *158*, 1330–1336. [CrossRef]

10. Patel, A.; Arora, N.; Mehtani, J.; Pruthi, V.; Pruthi, P.A. Assessment of fuel properties on the basis of fatty acid profiles of oleaginous yeast for potential biodiesel production. *Renew. Sustain. Energy Rev.* **2017**, *77*, 604–616. [CrossRef]

11. Papanikolaou, S.; Aggelis, G. Lipids of oleaginous yeasts. Part I: Biochemistry of single cell oil production. *Eur. J. Lipid Sci. Technol.* **2011**, *113*, 1031–1051. [CrossRef]

12. Papanikolaou, S.; Aggelis, G. Lipids of oleaginous yeasts. Part II: Technology and potential applications. *Eur. J. Lipid Sci. Technol.* **2011**, *113*, 1052–1073. [CrossRef]

13. Ban, J.; Arellano, J.L.; Alawami, A.; Aguilera, R.F.; Tallett, M. *2016 World Oil Outlook*, 10th ed.; Griffin, J., Fantini, A.-M., Eds.; OPEC's Economic Commission Board: Vienna, Austria, 2016; ISBN 978-3-9503936-2-0.

14. Ragauskas, A.J. The Path Forward for Biofuels and Biomaterials. *Science* **2006**, *311*, 484–489. [CrossRef] [PubMed]

15. Papanikolaou, S.; Aggelis, G. Lipid production by Yarrowia lipolytica growing on industrial glycerol in a single-stage continuous culture. *Bioresour. Technol.* **2002**, *82*, 43–49. [CrossRef]

16. Liang, Y.; Cui, Y.; Trushenski, J.; Blackburn, J.W. Converting crude glycerol derived from yellow grease to lipids through yeast fermentation. *Bioresour. Technol.* **2010**, *101*, 7581–7586. [CrossRef]

17. Xue, F.; Miao, J.; Zhang, X.; Luo, H.; Tan, T. Studies on lipid production by Rhodotorula glutinis fermentation using monosodium glutamate wastewater as culture medium. *Bioresour. Technol.* **2008**, *99*, 5923–5927. [CrossRef] [PubMed]

18. Patel, A.; Pruthi, V.; Pruthi, P.A. Synchronized nutrient stress conditions trigger the diversion of CDP-DG pathway of phospholipids synthesis towards de novo TAG synthesis in oleaginous yeast escalating biodiesel production. *Energy* **2017**, *139*, 962–974. [CrossRef]

19. Beopoulos, A.; Chardot, T.; Nicaud, J.-M.M. Yarrowia lipolytica: A model and a tool to understand the mechanisms implicated in lipid accumulation. *Biochimie* **2009**, *91*, 692–696. [CrossRef]

20. Fickers, P.; Benetti, P.H.; Waché, Y.; Marty, A.; Mauersberger, S.; Smit, M.S.; Nicaud, J.M. Hydrophobic substrate utilisation by the yeast Yarrowia lipolytica, and its potential applications. *FEMS Yeast Res.* **2005**, *5*, 527–543. [CrossRef]

21. Thevenieau, F.; Beopoulos, A.; Desfougeres, T.; Sabirova, J.; Albertin, K.; Zinjarde, S.; Nicaud, J. Uptake and Assimilation of Hydrophobic Substrates by the Oleaginous Yeast Yarrowia lipolytica. *Handb. Hydrocarb. Lipid Microbiol.* **2010**, *1*, 1513–1661. [CrossRef]

22. Käppeli, O.; Walther, P.; Mueller, M.; Fiechter, A. Structure of the cell surface of the yeast Candida tropicalis and its relation to hydrocarbon transport. *Arch. Microbiol.* **1984**, *138*, 279–282. [CrossRef] [PubMed]

23. Carsanba, E.; Papanikolaou, S.; Erten, H. Production of oils and fats by oleaginous microorganisms with an emphasis given to the potential of the nonconventional yeast Yarrowia lipolytica. *Crit. Rev. Biotechnol.* **2018**, *38*, 1230–1243. [CrossRef] [PubMed]

24. Aggelis, G.; Papadiotis, G.; Komaitis, M. Microbial fatty acid specificity. *Folia Microbiol. (Praha).* **1997**, *42*, 117–120. [CrossRef] [PubMed]

25. Papanikolaou, S.; Aggelis, G. Modeling lipid accumulation and degradation in Yarrowia lipolytica cultivated on industrial fats. *Curr. Microbiol.* **2003**, *46*, 398–402. [CrossRef] [PubMed]

26. Cirigliano, M.C.; Carman, G.M. Isolation of a Bioemulsifier from Candida-Lipolytica. *Appl. Environ. Microbiol.* **1984**, *48*, 747–750. [PubMed]

27. Martínez, E.J.; Raghavan, V.; González-Andrés, F.; Gómez, X. New biofuel alternatives: Integrating waste management and single cell oil production. *Int. J. Mol. Sci.* **2015**, *16*, 9385–9405. [CrossRef] [PubMed]

28. Anuar, M.R.; Abdullah, A.Z. Challenges in biodiesel industry with regards to feedstock, environmental, social and sustainability issues: A critical review. *Renew. Sustain. Energy Rev.* **2016**, *58*, 208–223. [CrossRef]
29. Kumar, M.; Pandya, H.B.; Dodiya, K.K.; Bhatt, R. Advancement in industrial method of ghee making process at Sarvottam Dairy, Bhavnagar, Gujarat (India). *Int. J. Sci. Environ.* **2017**, *6*, 1727–1736.
30. Khanam, R.; Prasuna, R.G. Comparison of extraction methods and solvents for total phenolics from dairy waste. *Asian J. Dairy Food Res.* **2017**, *36*, 251–255. [CrossRef]
31. Sahasrabudhe, J.; Palshikar, S.; Goja, A.; Kulkarni, C. Use of Ghee Residue as a Substrate for Microbial Lipase Production. *Int. J. Sci. Technol. Res.* **2012**, *1*, 61–64.
32. Khanam, R.; Prasuna, R.G.; Akbar, S. Exploitation of Industrial Waste Ghee Residue in Laccases Production by Pycnoporus cinnabarinus strain SYBC-L14. *Int. J. Plant Anim. Environ. Sci.* **2012**, *2*, 1–10.
33. Chkraboi, B.K. Industrial Ghee Production TrendS and Innovation. *Indian Dairym.* **1980**, *32*, 737–742.
34. Khanam, R.; Prasuna, R.G.; Akbar, S. Evaluation of Total Phenolic Content in Ghee Residue: Contribution to Higher Laccase Production. *Microbiol. J.* **2013**, *3*, 12–20. [CrossRef]
35. Poopathi, S.; Abibidha, S. The use of clarified butter sediment waste from dairy industries for the production of mosquitocidal bacteria. *Int. J. Dairy Technol.* **2012**, *65*, 152–157. [CrossRef]
36. Patel, A.; Pravez, M.; Deeba, F.; Pruthi, V.; Singh, R.P.; Pruthi, P.A. Boosting accumulation of neutral lipids in Rhodosporidium kratochvilovae HIMPA1 grown on hemp (Cannabis sativa Linn) seed aqueous extract as feedstock for biodiesel production. *Bioresour. Technol.* **2014**, *165*, 214–222. [CrossRef] [PubMed]
37. Patel, A.; Pruthi, V.; Singh, R.P.; Pruthi, P.A. Synergistic effect of fermentable and non-fermentable carbon sources enhances TAG accumulation in oleaginous yeast Rhodosporidium kratochvilovae HIMPA1. *Bioresour. Technol.* **2015**, *188*, 136–144. [CrossRef] [PubMed]
38. Patel, A.; Matsakas, L.; Rova, U.; Christakopoulos, P. Heterotrophic cultivation of Auxenochlorella protothecoides using forest biomass as a feedstock for sustainable biodiesel production. *Biotechnol. Biofuels* **2018**, *11*, 169. [CrossRef] [PubMed]
39. Talebi, A.F.; Tabatabaei, M.; Chisti, Y. BiodieselAnalyzer: A User-Friendly Software for Predicting the Properties of Prospective Biodiesel. *Biofuel Res. J.* **2014**, *2*, 55–57. [CrossRef]
40. Beopoulos, A.; Nicaud, J.-M.; Gaillardin, C. An overview of lipid metabolism in yeasts and its impact on biotechnological processes. *Appl. Microbiol. Biotechnol.* **2011**, *90*, 1193–1206. [CrossRef]
41. Thevenieau, F.; Le Dall, M.T.; Nthangeni, B.; Mauersberger, S.; Marchal, R.; Nicaud, J.M. Characterization of Yarrowia lipolytica mutants affected in hydrophobic substrate utilization. *Fungal Genet. Biol.* **2007**, *44*, 531–542. [CrossRef]
42. Garti, N.; Yaghmur, A.; Leser, M.E.; Clement, V.; Watzke, H.J. Improved oil solubilization in oil/water food grade microemulsions in the presence of polyols and ethanol. *J. Agric. Food Chem.* **2001**, *49*, 2552–2562. [CrossRef] [PubMed]
43. Huang, C.; Luo, M.T.; Chen, X.F.; Qi, G.X.; Xiong, L.; Lin, X.Q.; Wang, C.; Li, H.L.; Chen, X. De Combined 'de novo' and 'ex novo' lipid fermentation in a mix-medium of corncob acid hydrolysate and soybean oil by Trichosporon dermatis. *Biotechnol. Biofuels* **2017**, *10*, 147. [CrossRef] [PubMed]
44. Patel, A.; Matsakas, L. A comparative study on de novo and ex novo lipid fermentation by oleaginous yeast using glucose and sonicated waste cooking oil. *Ultrason. Sonochem.* **2019**, *52*, 364–374. [CrossRef]
45. Mlícková, K.; Roux, E.; Athenstaedt, K.; D'Andrea, S.; Daum, G.; Chardot, T.; Nicaud, J. Lipid accumulation, lipid body formation, and acyl coenzyme A oxidases of the yeast Yarrowia lipolytica. *Appl. Environ. Microbiol.* **2004**, *70*, 3918–3924. [CrossRef] [PubMed]
46. Papanikolaou, S.; Dimou, A.; Fakas, S.; Diamantopoulou, P.; Philippoussis, A.; Galiotou-Panayotou, M.; Aggelis, G. Biotechnological conversion of waste cooking olive oil into lipid-rich biomass using Aspergillus and Penicillium strains. *J. Appl. Microbiol.* **2011**, *110*, 1138–1150. [CrossRef]
47. Saenge, C.; Cheirsilp, B.; Tachapattaweawrakul, T.; Bourtoom, T.; Suksaroge, T.T.; Bourtoom, T. Potential use of oleaginous red yeast Rhodotorula glutinis for the bioconversion of crude glycerol from biodiesel plant to lipids and carotenoids. *Process Biochem.* **2011**, *46*, 210–218. [CrossRef]
48. Papanikolaou, S.; Aggelis, G. Biotechnological valorization of biodiesel derived glycerol waste through production of single cell oil and citric acid by Yarrowia lipolytica. *Lipid Technol.* **2009**, *21*, 83–87. [CrossRef]
49. Aggelis, G.; Komaitis, M.; Papanikolaou, S.; Papadopoulos, G.; Papadopoulos, G. A mathematical model for the study of lipid accumulation in oleaginous microorganisms. I. Lipid accumulation during growth of Mucor circinelloides CBS 172-27 on a vegetable oil. *Grasas y Aceites* **1995**, *46*, 169–1873. [CrossRef]

50. Boyle, N.R.; Morgan, J.A. Flux balance analysis of primary metabolism in Chlamydomonas reinhardtii. *BMC Syst. Biol.* **2009**, *3*, 1–14. [CrossRef]

51. Wu, S.; Zhao, X.; Shen, H.; Wang, Q.; Zhao, Z.K. Microbial lipid production by Rhodosporidium toruloides under sulfate-limited conditions. *Bioresour. Technol.* **2011**, *102*, 1803–1807. [CrossRef]

52. Wu, S.; Hu, C.; Jin, G.; Zhao, X.; Zhao, Z.K. Phosphate-limitation mediated lipid production by Rhodosporidium toruloides. *Bioresour. Technol.* **2010**, *101*, 6124–6129. [CrossRef]

53. Jiru, T.M.; Steyn, L.; Pohl, C.; Abate, D. Production of single cell oil from cane molasses by Rhodotorula kratochvilovae (syn, Rhodosporidium kratochvilovae) SY89 as a biodiesel feedstock. *Chem. Cent. J.* **2018**, *12*, 1–7. [CrossRef] [PubMed]

54. Ratledge, C. Fatty acid biosynthesis in microorganisms being used for Single Cell Oil production. *Biochimie* **2004**, *86*, 807–815. [CrossRef]

55. Ratledge, C.; Wynn, J.P. The biochemistry and molecular biology of lipid accumulation in oleaginous microorganisms. *Adv. Appl. Microbiol.* **2002**, *51*, 1–51. [PubMed]

56. Papanikolaou, S. Oleaginous Yeasts: Biochemical Events Related with Lipid Synthesis and Potential Biotechnological Applications. *Ferment. Technol.* **2012**, *1*, e103. [CrossRef]

57. Guo, Y.; Cordes, K.R.; Farese, R.V.; Walther, T.C. Lipid droplets at a glance. *J. Cell Sci.* **2009**, *122*, 749–752. [CrossRef]

58. Guo, T.; Kit, Y.Y.; Nicaud, J.M.; Le Dall, M.T.; Sears, S.K.; Vali, H.; Chan, H.; Rachubinski, R.A.; Titorenko, V.I. Peroxisome division in the yeast Yarrowia lipolytica is regulated by a signal from inside the peroxisome. *J. Cell Biol.* **2003**, *162*, 1255–1266. [CrossRef]

59. Yen, H.W.; Hu, C.Y.; Liang, W.S. A cost efficient way to obtain lipid accumulation in the oleaginous yeast Rhodotorula glutinis using supplemental waste cooking oils (WCO). *J. Taiwan Inst. Chem. Eng.* **2019**, *97*, 80–87. [CrossRef]

60. Papanikolaou, I.; Chevalot, M.; Komai, S. Single cell oil production by Yarrowia lipolytica growing on an industrial derivative of animal fat in batch cultures. *Appl. Microbiol. Biotechnol.* **2002**, *58*, 308–312. [CrossRef]

61. Bati, N.; Hammond, E.G.; Glatz, B.A. Biomodification of fats and oils: Trials with Candida lipolytica. *J. Am. Oil Chem. Soc.* **1984**, *61*, 1743–1746. [CrossRef]

62. Katre, G.; Ajmera, N.; Zinjarde, S.; RaviKumar, A. Mutants of Yarrowia lipolytica NCIM 3589 grown on waste cooking oil as a biofactory for biodiesel production. *Microb. Cell Fact.* **2017**, *16*, 176. [CrossRef] [PubMed]

63. Martin, S.; Parton, R.G. Lipid droplets: A unified view of a dynamic organelle. *Nat. Rev. Mol. Cell Biol.* **2006**, *7*, 373–378. [CrossRef] [PubMed]

64. Ploegh, H.L. A lipid-based model for the creation of an escape hatch from the endoplasmic reticulum. *Nature* **2007**, *448*, 435–438. [CrossRef] [PubMed]

65. Kohlwein, S.D.; Veenhuis, M.; van der Klei, I.J. Lipid droplets and peroxisomes: Key players in cellular lipid homeostasis or a matter of fat—Store 'em up or burn 'em down. *Genetics* **2013**, *193*, 1–50. [CrossRef] [PubMed]

66. Patel, A.; Arora, N.; Pruthi, V.; Pruthi, P.A. Biological treatment of pulp and paper industry effluent by oleaginous yeast integrated with production of biodiesel as sustainable transportation fuel. *J. Clean. Prod.* **2017**, *142*, 2858–2864. [CrossRef]

67. Lanjekar, R.D.; Deshmukh, D. A review of the effect of the composition of biodiesel on NOx emission, oxidative stability and cold flow properties. *Renew. Sustain. Energy Rev.* **2016**, *54*, 1401–1411. [CrossRef]

68. Hoekman, S.K.; Broch, A.; Robbins, C.; Ceniceros, E.; Natarajan, M. Review of biodiesel composition, properties, and specifications. *Renew. Sustain. Energy Rev* **2012**, *16*, 143–169. [CrossRef]

69. Kumar, M.; Sharma, M.P. Selection of potential oils for biodiesel production. *Renew. Sustain. Energy Rev.* **2016**, *56*, 1129–1138. [CrossRef]

70. Kumar, N. Oxidative stability of biodiesel: Causes, effects and prevention. *Fuel* **2017**, *190*, 328–350. [CrossRef]

71. Ramírez-Verduzco, L.F.; Rodríguez-Rodríguez, J.E.; Jaramillo-Jacob, A.D.R. Predicting cetane number, kinematic viscosity, density and higher heating value of biodiesel from its fatty acid methyl ester composition. *Fuel* **2012**, *91*, 102–111. [CrossRef]

72. Oliveira, L.E.; Da Silva, M.L.C.P. Relationship between cetane number and calorific value of biodiesel from Tilapia visceral oil blends with mineral diesel. In Proceedings of the International Conference on Renewable Energies and Power Quality (ICREPQ'13), Bilbao, Spain, 20–23 March 2013.

73. Knothe, G.; Razon, L.F. Biodiesel fuels. *Prog. Energy Combust. Sci.* **2017**, *58*, 36–59. [CrossRef]

Foods **2019**, *8*, 234

74. Mahmudul, H.M.; Hagos, F.Y.; Mamat, R.; Adam, A.A.; Ishak, W.F.W.; Alenezi, R. Production, characterization and performance of biodiesel as an alternative fuel in diesel engines—A review. *Renew. Sustain. Energy Rev.* **2017**, *72*, 497–509. [CrossRef]

75. Suh, H.K.; Lee, C.S. A review on atomization and exhaust emissions of a biodiesel-fueled compression ignition engine. *Renew. Sustain. Energy Rev.* **2016**, *58*, 1601–1620. [CrossRef]

Article

How Do Arabinoxylan Films Interact with Water and Soil?

Cassie Anderson and Senay Simsek *

North Dakota State University, Department of Plant Sciences, Cereal Science Graduate Program, Fargo, ND 6050, USA; Cassie.anderson@ndsu.edu
* Correspondence: senay.simsek@ndsu.edu; Tel.: +701-231-7737

Received: 28 May 2019; Accepted: 12 June 2019; Published: 17 June 2019

Abstract: Biodegradable materials made from cereal arabinoxylan could provide an alternative source of packaging to replace current nonbiodegradable plastics. The main purpose of this research was to determine how arabinoxylan (AX) films made from wheat bran (WB) AX, maize bran (MB) AX, and dried distillers grain (DDG) AX made with either glycerol or sorbitol at varying levels (10, 25 or 50%) interacts with soil and water. The biodegradability of all films ranged from 49.4% biodegradable (DDG AX with 10% sorbitol) to 67.7% biodegradable (MB AX with 50% glycerol). In addition, the MB AX films with 25% sorbitol had the lowest moisture content at 9.7%, the MB AX films with 10% glycerol had the highest water solubility at 95.6%, and the MB AX films with 50% glycerol had the highest water vapor transmission rate (WVTR) at 90.8 g h^{-1} m^{-2}. Despite these extreme trends in the MB AX films, the WB AX films were the least hydrophilic on average while the DDG AX films were the most hydrophilic on average. The 18 materials developed in this research demonstrate varying affinities for water and biodegradation. These materials can be used for many different packaging materials, based on their unique characteristics.

Keywords: arabinoxylan; films; biodegradability; hydrophilicity; food packaging; sustainability

---

1. Introduction

Wheat (*Triticum aestivum*) and maize (*Zea mays*) are members of the *Gramineae* family and are two of the top three most commonly produced cereal crops [1,2]. The fruit of these cereals is known as a caryopsis and consists mainly of starch, protein, and non-starch polysaccharides [1,3]. The outer portion of a caryopsis is the bran, which makes up 5 to 14% of the caryopsis depending upon the crop [1,4]. The wheat and maize bran is often discarded as a byproduct of milling as it is not desirable for all products (e.g., refined flour). In addition to bran, the dried distillers grain (DDG) is a byproduct of processing maize into ethanol.

One of the main components in these three byproducts is arabinoxylan (AX). It is the main type of non-starch polysaccharide present in wheat bran (WB), maize bran (MB), and DDG. Arabinoxylan is present in both wheat and maize cell walls, and it is one of the most common non-starch polysaccharides found on earth [5]. It is made up of a backbone of β-1,4 linked xylose units that have O-2 and/or O-3 linked arabinose substituents [5–7]. In addition to AX, two other polysaccharides commonly present in AX are glucose and ferulic acid [8–10]. Ferulic acid can be substituted on the xylose backbone in the O-3 and/or O-2 locations [6,11,12]. Arabinoxylan cross-linkages are often formed in the presence of ferulic acid because ferulic acid can couple at the O-5 location via an ester linkage.

Arabinoxylan can be extracted from WB, MB, and DDG through a variety of methods including alkaline extraction [13]. This AX can then be used to make films, which are utilized as a prototype of food packaging when testing material properties. Materials used for food packaging must have adequate mechanical properties and extend the shelf-life of the food being stored. Due to this, AX is often combined with a plasticizer, such as glycerol or sorbitol, to increase both flexibility and strength [14,15].

Food packaging has historically been created from synthetic materials including polyolefins, polyamides, and polyesters [16,17]. These materials have been used because of their desirable barrier properties that assist in elongating the shelf life of the food(s) they were packaging. However, the use of synthetic packaging materials, that are not biodegradable or recyclable, has resulted in many ecological problems. When developing food packaging for future use, biodegradability is an important factor to consider. Utilization of AX from WB, MB, and DDG for use as the basis in food packaging materials will increase the overall sustainability of the food packaging industry because these materials are biodegradable [17].

The objectives of this research were to utilize AX of high purity to create film materials, to determine how the films interact with both water and soil (in aerobic biodegradation), and to determine if these interactions are correlated to the physicochemical properties of the films.

2. Materials and Methods

2.1. Arabinoxylan Preparation and Characterization

The WB, MB, and DDG utilized in this research were provided by the North Dakota State Mill (Grand Forks, ND, USA), Agricor, Inc. (Marion, Indiana), and Tharaldson Ethanol (Casselton, ND, USA), respectively. These materials were commercially available, standard materials produced by each manufacturer.

The WB AX, MB AX, and DDG AX used as the basis of the films in this research was alkaline extracted from the three starting materials and subsequently purified using a standard method [18]. The moisture content, protein content, ash content, AX content, Arabinose to Xylose (A/X) ratio, M_w, polydispersity index, and linkages of the three types of AX were also determined using standard methods [19–23].

The AX content and A/X ratio were determined by derivatizing the samples to alditol acetates and analysis with gas chromatography with flame ionization detection [23]. The gas chromatograph used to determine the AX content was a Hewlett Packard 5890 Series II GC system with a flame ionization detector (Agilent Technologies, Incorporated Santa Clara, CA). The column used was a SupelcoSP-2380 fused silica capillary column (30 m × 0.25 mm × 2 μm) (Supelco Bellefonte, PA). The flow pressure was 0.83 MPa, the oven temperature was 100 °C, the flow rate was 0.8 mL min^{-1}, the detector temperature was 250 °C, the injector temperature of 230 °C, and the carrier gas was He. The AX content and A/X ratio were calculated by the formulas:

$$\% \text{ AX} = [((\text{Arabinose} + \text{Xylose})*0.88) \div (1000*\text{Weight})]*100 \tag{1}$$

$$\text{A/X ratio} = \text{Arabinose} \div \text{Xylose} \tag{2}$$

To determine the M_w and polydispersity index of the AX, a high performance liquid chromatograph (HPLC) with a multi-angle light scattering detector and a refractive index detector were used. The HPLC used was an Agilent 1200 with a Wyatt Dawn Helios-II multi-angle light scattering detector and a refractive index detector (Agilent Technologies, Incorporated Santa Clara, CA, USA). Two columns were used during analysis: A Shodex OHpak guard column and an SB 806-HQ column. The flow rate during analysis was 0.5 mL min^{-1}. The Astra 6.0.5 software and a 3rd order Debye plot with the second-order polynomial fit was used for data analysis [24] The proportional change in the refractive index with changes in polymer concentration for AX were assumed to be 0.146 as previously published in research by Dervilly et al. [25].

For AX linkage assessment, the nuclear magnetic resonance spectrometer (NMR) utilized was a 400 MHz spectrometer (Brunker, Billerica, MA, USA) (Bruker AV3 HD 400 MHZ NMR that had a 5 mm PABBO BB/19F-!H/D Z-GRD Z probe). TopSpin 3.2 software (TopSpin, Billerica, MA, USA) was used to analyze all data [26].

2.2. Preparation of Arabinoxylan Films

All films were created according to the methods of Anderson and Simsek [18]. The sorbitol (BioUltra Grade) and glycerol (ACS Reagent Grade) were purchased from Sigma-Aldrich (Saint Louis, MO, USA). Film solutions were made by preparing a 26.7 g kg^{-1} solution of AX in deionized water. The solutions were stirred for 24 h then divided into aliquots and heated at 90 °C for 15 min. Then, sorbitol or glycerol was added (100, 250 or 500 g kg^{-1}) and the samples were vortexed. After heating at 90 °C for an additional 15 min, the samples were cast onto polystyrene Petri dishes and dried at 60 °C for 8 h. The films were allowed to finish drying overnight at 23 °C. All films were stored in a Dry-Keeper (Sanplatec Corporation, Catalog No. H42056-0001, Osaka, Japan) with Boveda 49% relative humidity packs (Item No. B49-60-48). Prior to the analysis of the interactions of the AX films with water, all films were conditioned at 23 °C and 50% relative humidity for at least 48 h prior to analyses.

2.3. Moisture Content Determination

The moisture content of each film was determined in triplicate following the method of Garcia et al. [27]. To begin, the initial masses of three films from each treatment were determined and recorded. Next, all films were placed into a 110 °C oven for 24 h. After 24 h, the final mass of each film was determined. The moisture content of each film was determined using the following equation:

$$\text{Film moisture content} = [(\text{Initial film mass} - \text{Final film mass}) \div (\text{Initial film mass})]*100 \qquad (3)$$

2.4. Water Solubility Analysis.

The percentage of water-soluble material in each film was determined in duplicate following a modified method of Garcia et al. [27]. To begin, each film was carefully cut into a 2 cm × 3 cm piece and the initial mass was recorded. Next, each film was placed into 80 mL of distilled water in a capped glass jar. The jars were then placed onto a Lab-Linee Orbit Shaker (Model No. 3520) and shaken at 75 rpm for one h at 25 °C. After one h, each sample was filtered through an Endecotts Limited (London, England) Steel Mesh No. 325 (Aperture: 45 microns). The solid material remaining was dried at 100 °C for 10 h and the final mass of the material was determined. The percentage of water-soluble material was determined using the following equation:

$$\text{Water soluble material} = [(\text{Initial film mass} - \text{Final film mass}) \div (\text{Initial film mass})]*100 \qquad (4)$$

2.5. Contact Angle and Wetting Tension Determination

The contact angle and wetting tension of all treatments was determined by following the American Society for Testing and Materials official method D7334-08 in duplicate for distilled water [28]. A Dynamic Contact Angle Analyzer by First Ten Angstroms 125 (Serial No. 980806, First Ten Angstroms, Portsmouth, VA, USA) with a charge coupled device camera was used for testing and calculation of contact angle and wetting tension. A drop of distilled water was applied to the film with a syringe and the contact angle of the drop of water with the film was measured using the dynamic contact angle analyzer instrument [28].

2.6. Water Vapor Permeability Analysis

The (WVTR) and permeance are part of the water vapor permeability measurement that was determined according to the American Society for Testing and Materials official method E96/E96M-15 in triplicate [29]. The average thickness of the films was 0.0792 mm. The water method was used, and the relative humidity and temperature were recorded every 30 min. During testing, the mass of each sample was determined using a Mettler Toledo New Classic MF analytical balance (Model No. ML203E/03). The test assembly was as follows: A polystyrene dish (100 mm in diameter) with 20 mL distilled water, two steel metal plates with inner diameters of 5.4 cm and outer diameters of 11.43 cm with plain finishes (Grainger Item No. 22UE14; Model No. U38402.200.0001) were used to hold the specimen flat, and parafilm wax was used to secure the specimen and seal the entire apparatus. The film specimens were 13 mm above the surface of the water. The test was conducted at room temperature, which was 22–25 °C, at the time of the test. The following equations were used to determine the water vapor transmission rate and water vapor permeance, respectively.

$$\text{Water vapor transmission rate} = G \div tA \tag{5}$$

where G = mass change (in grams), t = time in h, $G\,t^{-1}$ = slope of a straight line in grams per h, A = test area in square meters.

$$\text{Water vapor permeance} = \text{WVTR} \div \Delta p = \text{WVTR} \div [S(R_1 - R_2)] \tag{6}$$

where WVTR = water vapor transmission rate, Δp = vapor pressure difference in mm Hg, S = saturation vapor pressure at test temperature in mm Hg, R_1 = relative humidity at the source, R_2 = relative humidity at the vapor sink. In this study $R_1 = 1.0$ and $R_2 = 0.26625$.

2.7. Biodegradability Analysis

The C content of all films was determined using a Primacs TOC Analyzer (Model CS22). The biodegradability analysis of the AX films was completed in triplicate according to a combination of the method of Colussi et al. [30] and the American Society for Testing and Materials official method D5988-12 [31]. Soil collected from Foster County, ND (100 g) was mixed with enough water to reach 60% of the moisture holding capacity (MHC) of the soil (60% MHC = 36.67g/100g soil), and half of the wetted soil was added to a 1-l airtight glass jar. Then, a 400 mg sample of AX film was placed on top of the soil and the film was covered with the remaining soil. A polystyrene cup (50 mL) with 20 mL of 1M NaOH was placed on top of the soil and the jar was sealed tightly. A control sample of soil only was used as a blank, and all samples were stored at room temperature (23 to 26 °C). The CO_2 production was measured over 145 days. The cups containing NaOH were removed from the jars on days 19, 40, 55, 82, 103, 119 and 145 and replaced with new cups containing fresh NaOH solution. The NaOH was titrated to determine the CO_2 produced. This titration was conducted by transferring the NaOH that had been stored with each film sample to a flask and adding 5 mL of $BaCl_2$ (25%). Three drops of 0.1% phenolphthalein were added and 1M HCl was used to titrate the solution until it turned from pink to white. The following equations were utilized to determine the total amount of biodegradable material in each type of film.

$$\text{Release of C per 100 g } CO_2 = (1 \div \text{mg } CO_2) - (AB)*(\text{Acid molarity})*(\text{Eq.g } C\text{-}CO_2) \tag{7}$$

where A is the amount of HCl spent in reagent blank in mL, B is the amount of hydrochloric acid spent in the sample in mL, acid molarity is the molarity of the HCl in M, and Eq.g. C-CO_2 is the equivalent gram C-CO_2.

$$\text{Percentage of film biodegraded} = [(CO_2 \text{ soil with film} - CO_2 \text{ soil without film}) \div (\text{mg C in film})]*100 \tag{8}$$

The soil (silt loam: 25.8% sand, 57.8% silt, 16.4% clay) used for testing was collected from Foster County, ND June of 2015. The sampling depth of the soil collected was 0 to 15 cm. It was stored at 23 °C in a closed container and handled with gloved hands and clean tools. The particle size was of the soil was less than 2 mm. The moisture holding capacity of the soil was 36.67 g water per 100 g soil. The C:N of the soil was 11:1, and the pH was 6.2.

2.8. Statistical Analysis

The experiment utilized a completely random design with a factorial arrangement that was utilized with three factors (AX type, plasticizer type, and plasticizer level). Statistical Analysis Software version 9.3 was utilized for all data analysis [32]. All data was analyzed using ANOVA and Fischer's protected least significant difference (LSD). The Pearson's correlation coefficient was utilized to assess the relationships between the five film properties and the physicochemical properties of the films.

3. Results and Discussion

3.1. Arabinoxylan Characterization

The AX extracts utilized in this research were of a purity of at least 58% as shown in Table 1. In addition, the arabinose to xylose ratio for all extracts was 0.51. The MB AX had the highest M_w indicating that this extract had the largest AX polymers, whereas the DDG AX had the lowest M_w and polydispersity index indicating these AX polymers were the smallest and the most similar in M_w. The WB AX had the highest polydispersity index indicating the largest range for AX M_w out of the three extracts.

Table 1. Proximate compositions (dry weight basis) of alkaline extracted wheat bran arabinoxylan, maize bran arabinoxylan, and dried distillers grain arabinoxylan.

	Moisture (%)	Ash (%)	Protein (%)	Arabinoxylan (%)	Molecular Weight (g/mol)	Polydispersity index
Wheat bran arabinoxylan	12.62	8.55	13.62	72.94	7.12×10^6	1.59
Maize bran arabinoxylan	11.11	1.22	3.89	84.71	7.70×10^6	1.27
Dried distillers grain arabinoxylan	6.31	1.89	14.95	58.05	5.90×10^6	1.04
LSD ($p < 0.05$)	1.39	0.06	1.31	2.97	0.60×10^5	0.23
LSD ($p < 0.01$)	2.56	0.11	2.40	5.45	0.80×10^5	0.31

LSD: Least Significant Difference.

Figure 1 provides the abundance of each type of linkage present in the three AX extracts. The linkage analysis showed that there were significant ($p < 0.05$) variations in the structures of the three types extracted AX. The variation in the three types of AX extracts is most likely due to the differences in the starting material AX structure [33]. These differences in linkages influenced the interactions of films made from these three types of AX with both water and soil in aerobic biodegradation.

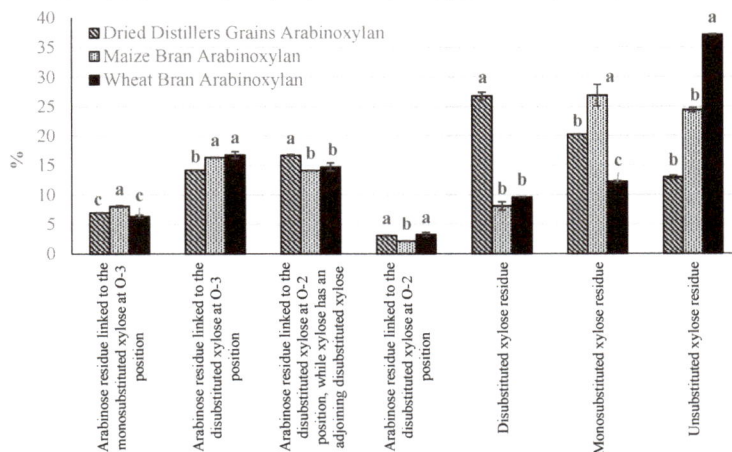

Figure 1. Abundance of seven types of anomeric protons present in dried distillers grain arabinoxylan, maize bran arabinoxylan, and wheat bran arabinoxylan as determined by [1]H Nuclear Magnetic Resonance Spectrometry. Columns with the same letter for the same proton are not significantly ($p < 0.05$) different, error bars represent standard deviation.

3.2. Interactions of Arabinoxylan Films with Water

Each of the 18 materials developed in this research interacted with water in different ways as shown in Tables 2 and 3. The moisture content, water solubility, hydrophilicity (as determined by contact angle), and water vapor permeability are all vital characteristics to food packaging materials. These properties must be carefully characterized so that a proper material can be developed to store food. When the proper packaging material is utilized, it can extend the shelf life of food, which may decrease food waste and increase sustainability in the food industry.

Table 2. Interactions of wheat bran arabinoxylan, maize bran arabinoxylan, and dried distillers grain arabinoxylan films with water.

Film Composition	Moisture Content	Water Soluble Material	Contact Angle	Wetting Tension
	(%)	(%)	(°)	(mN m^{-1})
WB AX [a] + 10% sorbitol	11.0 ± 1.9	54.7 ± 0.4	62.81 ± 0.66	33.27 ± 0.75
WB AX + 25% sorbitol	11.1 ± 0.3	36.0 ± 2.3	63.97 ± 1.68	31.94 ± 1.91
WB AX + 50% sorbitol	11.6 ± 0.2	44.0 ± 3.3	66.59 ± 0.34	30.02 ± 1.15
WB AX + 10% glycerol	12.1 ± 0.8	38.0 ± 4.7	76.71 ± 0.68	17.02 ± 1.24
WB AX + 25% glycerol	16.0 ± 0.5	30.5 ± 0.1	101.30 ± 0.47	−14.26 ± 0.58
WB AX + 50% glycerol	29.0 ± 0.7	34.2 ± 1.8	102.01 ± 0.39	−15.15 ± 0.49
MB AX [b] + 10% sorbitol	11.0 ± 0.3	87.9 ± 0.3	71.03 ± 0.78	23.66 ± 0.93
MB AX + 25% sorbitol	9.7 ± 0.1	86.3 ± 6.6	71.52 ± 0.69	23.08 ± 0.84
MB AX + 50% sorbitol	10.2 ± 0.3	70.5 ± 6.7	74.25 ± 1.04	19.77 ± 1.27
MB AX + 10% glycerol	10.3 ± 0.6	95.6 ± 3.0	71.10 ± 0.66	23.58 ± 0.79
MB AX + 25% glycerol	19.3 ± 0.9	71.7 ± 1.4	82.25 ± 3.00	9.81 ± 3.77
MB AX + 50% glycerol	21.3 ± 1.4	62.5 ± 5.5	64.45 ± 0.51	31.40 ± 0.58
DDG AX [c] + 10% sorbitol	10.5 ± 0.5	93.7 ± 2.1	54.23 ± 2.67	42.54 ± 2.75
DDG AX + 25% sorbitol	10.2 ± 1.3	87.3 ± 1.1	52.49 ± 0.01	44.33 ± 0.01
DDG AX + 50% sorbitol	10.6 ± 0.4	89.3 ± 3.2	62.29 ± 0.75	33.85 ± 0.85
DDG AX + 10% glycerol	13.4 ± 0.8	82.6 ± 0.8	61.27 ± 3.56	34.96 ± 3.97
DDG AX + 25% glycerol	11.9 ± 0.1	84.5 ± 5.9	46.67 ± 1.21	49.96 ± 1.12
DDG AX + 50% glycerol	37.1 ± 0.7	73.2 ± 6.6	53.88 ± 0.61	42.91 ± 0.62
LSD ($p < 0.05$)	1.3	8.1	3.06	3.56
LSD ($p < 0.01$)	1.8	11.1	4.20	4.88

Results are expressed as means ± standard deviation. [a] Wheat bran arabinoxylan; [b] maize bran arabinoxylan; [c] dried distillers grain arabinoxylan. AX: arabinoxylan; WB: wheat bran; MB: maize bran; DDG: dried distillers grain.

When the type of AX (WB, MB, or DDG) is considered, there are clear trends in the interaction of these AX films with water. Taking all four measurements of the interactions of these 18 types of AX films with water into account, the films made with WB AX appear to be the least hydrophilic and the DDG AX films appear to be the most hydrophilic. First, there were significant ($p < 0.05$) differences in the moisture content of the films. The films with the lowest moisture content were the MB AX films (average moisture content of 13.7%), while DDG AX films had the highest moisture contents (average moisture content of 15.6%). There were also significant ($p < 0.05$) increases in moisture content for films with increasing levels of glycerol. There was a significant ($p < 0.05$) correlation between the polydispersity index of the AX in the films and their moisture content for films made with glycerol ($r = 0.77$).

As these films became more heterogeneous in M_w they became more susceptible to moisture penetration. Second, the DDG AX films had the highest amount of water-soluble material (averaging 85.1%), followed by the MB AX films (averaging 79.1%), and lastly the WB AX films (averaging 39.5%). There were significant ($p < 0.05$) decreases in water solubility when films were prepared with higher levels of the sorbitol or glycerol plasticizers. There was a significant ($p < 0.01$) correlation between the M_w of the films and their water solubility ($r = -0.78$ for films made with sorbitol and $r = -0.91$ for films made with glycerol). In addition, there were significant ($p < 0.05$) correlations between an increase in the presence of disubstituted xylose and an increase in the amount of water-soluble material present in the film correlations ($r = 0.76$ for films made with sorbitol and $r = 0.67$ for films made with glycerol). This was due to a decrease in inter-polymer interaction that allowed water to enter and dissolve the material. Third, the WB AX films were the least hydrophilic with average contact angles of 79° and wetting tensions of 13.8 mN m^{-1}. The MB AX films were the next hydrophilic with average contact angles of 72° and average wetting tensions of 21.9 mN m^{-1} on average. In general, films made with sorbitol had significantly ($p < 0.05$) lower contact angle than films made with sorbitol, but films made with sorbitol tended to have significantly ($p < 0.05$) higher wetting tension than those prepared with glycerol. The average contact angles and average wetting tensions of the DDG AX films were 55° and 41.4 mN m^{-1}, respectively. There was a significant ($p < 0.01$) correlation between the presence of unsubstituted xylose on the AX polymer used in the films and the contact angle of the films with water ($r = 0.81$ for films made with sorbitol and $r = 0.74$ for films made with glycerol) demonstrating that as the branching on the AX polymer increases, the films became more hydrophilic. Fourth, there were significant ($p < 0.05$) differences in WVTR and water permeance among the films. The MB AX films were the most permeable to water vapor with an average WVTR of 64.6 g h^{-1}m^{-2} and average permeance of 411.0 g/s·m^2·Pa. The films made with WB AX were the least permeable to water vapor and had an average WVTR of 54.7 g h^{-1}m^{-2} and average permeance of 348.7 g/s·m^2·Pa. The level of sorbitol did not significantly ($p < 0.05$) affect the WVTR for films prepared from WB AX or MB AX, but the WVTR was significantly ($p < 0.05$) higher in DDG AX films prepared with 50% sorbitol than the DDG AX films with 10% or 25% sorbitol. For the most part, films made with glycerol had significantly ($p < 0.05$) higher WVTR and water permeance than films made with sorbitol.

Table 3. Water vapor transmission rate and water permeance of wheat bran arabinoxylan, maize bran arabinoxylan, and dried distillers grain arabinoxylan films.

Film Composition	Water Vapor Transmission Rate	Water Permeance
	(g h^{-1} m^{-2})	(g/s·m^2·Pa)
WB AX [a] + 10% sorbitol	44.8 ± 3.1	285.4 ± 19.5
WB AX + 25% sorbitol	45.4 ± 0.9	289.6 ± 5.6
WB AX + 50% sorbitol	47.3 ± 1.8	301.6 ± 11.8
WB AX + 10% glycerol	56.6 ± 3.8	360.4 ± 24.4
WB AX + 25% glycerol	60.6 ± 7.0	386.0 ± 44.6
WB AX + 50% glycerol	73.7 ± 3.6	469.4 ± 22.6
MB AX [b] + 10% sorbitol	50.6 ± 1.8	321.5 ± 11.8
MB AX + 25% sorbitol	53.2 ± 2.9	338.3 ± 18.5
MB AX + 50% sorbitol	54.1 ± 0.8	344.0 ± 5.0
MB AX + 10% glycerol	60.1 ± 1.1	382.4 ± 6.9
MB AX + 25% glycerol	78.9 ± 3.6	502.0 ± 23.2
MB AX + 50% glycerol	90.8 ± 1.6	577.8 ± 10.3
DDG AX [c] + 10% sorbitol	48.5 ± 3.1	308.8 ± 19.5
DDG AX + 25% sorbitol	49.8 ± 0.8	316.8 ± 5.0
DDG AX+ 50% sorbitol	55.16 ± 4.45	350.8 ± 28.3
DDG AX + 10% glycerol	51.52 ± 1.60	328.2 ± 10.2
DDG AX + 25% glycerol	60.95 ± 2.72	388.3 ± 17.3
DDG AX + 50% glycerol	78.04 ± 2.46	497.2 ± 15.7
LSD ($p < 0.05$)	5.02	32.0
LSD ($p < 0.01$)	6.73	42.9

Results are expressed as means ± standard deviation. [a] Wheat bran arabinoxylan; [b] maize bran arabinoxylan; [c] dried distillers grain arabinoxylan.

The interaction of each type of AX film with water depended not only on the type of AX but also on the type of plasticizer present in the film. Glycerol is more hydrophilic than sorbitol and has a greater plasticizing ability [34]. The films developed in this research that were made with glycerol had an average moisture content of 18.9%, while those made with sorbitol had an average moisture content of 11%. This trend was also noted in previous work published by Antoniou et al. [35].

In addition, films made with sorbitol were significantly ($p < 0.01$) more water soluble (with an average of 72.2% water-soluble material) than those made with glycerol (with an average of 63.7% water-soluble material). This result is the same as that found in films made with varying levels of sorbitol or glycerol by Müller et al. [34]. Similarly, films made with glycerol had contact angles of 73° and wetting tensions of 20.0 mN m^{-1} on average, whereas films made with sorbitol had contact angles of 64° and wetting tensions of 31.4 mN m^{-1} on average. Lastly, the films made with sorbitol had a significantly ($p < 0.01$) lower WVTR than those made with glycerol. The films made with sorbitol had an average WVTR of 49.9 g h^{-1}m^{-2} and average permeance of 317.4 g/s·m^2·Pa. The films made with glycerol had an average WVTR of 67.9 g h^{-1}m^{-2} and average permeance of 432.4 g/s·m^2·Pa. Zhang and Whistler also observed this trend in the WVTR of their MB AX films [15].

The amount of plasticizer also played a major role in the interactions of these AX films with water. The moisture contents of the films increased significantly ($p < 0.01$) from 11.4 to 20.0% as the level of plasticizer increased from 10 to 50%. However, there was a decrease in water-soluble material present in the films as the plasticizer level increased. The films made with 10% plasticizer had 75.4% water-soluble material on average, whereas those made with 25% plasticizer or 50% plasticizer had 66.1% water-soluble material or 62.3% water-soluble material, respectively. This trend was also identified by Nazan Turhan and Şahbaz in the films they created [36]. This is most likely due to the increase in intermolecular interactions due to the presence of the plasticizer, which limits the interaction of the AX polymers with water. In addition, as the level of plasticizer increased, there was a significant ($p < 0.05$) decrease in the hydrophilicity of the films as measured by contact angle and wetting tension. As the plasticizer content increased from 10 to 50%, the contact angles of the films increased (on average) from 66° to 71°, respectively. Similarly, the wetting tensions of the same films decreased (on average) from 29 to 24 mN m^{-1} as the plasticizer content increased from 10 to 50%, respectively. These results show the same trend as those of previously published work with variation in plasticizer levels and surface hydrophilicity [37]. Finally, as the plasticizer level in the films increased from 10 to 25 to 50%, the average WVTR increased from 52.0 to 58.2 to 66.5 g h^{-1}m^{-2}, respectively. In addition, the average permeance also increased from 331.1 to 370.2 to 423.5 g/s·m^2·Pa as the plasticizer level in the films increased from 10 to 25 to 50%, respectively. This is the same trend observed by Nazan Turhan and Şahbaz in their research on water vapor permeability of films at varying plasticizer levels [36]. These increases in water vapor permeability could be due to the decreased hydrophilicity of the films and the decreased order of the AX polymers allowing more water vapor to pass through.

3.3. Arabinoxylan Film Biodegradability

While food packaging must be strong enough to protect the food it contains, it is also beneficial if it is biodegradable. The biodegradation of bio-based materials such as AX films will vary from material to material, but the microorganisms in soil capable of breaking down AX are those that can produce hemicellulases such as *Bacillus* spp. [38,39]. The biodegradability of the materials developed in this research is given in Table 4. All 18 types of AX film rapidly biodegraded within the first 45 days of measurement followed by a plateau in biodegradation (Figure 2).

It has been documented that during this time, the components of plant-based materials are broken down sequentially [40]. Soluble carbohydrates are broken down first followed by proteins and structural carbohydrates. The hemicellulose and cellulose present are broken down last [40]. The biodegradability profiles for the AX films are given in Figure 2.

On average, the total amount of biodegradable material in each type of films was as follows: 53% in DDG AX films, 55% in WB AX films, and 63% in MB AX films. There was a significant ($p < 0.01$) correlation between the polydispersity index of the films and their biodegradability ($r = -0.70$ for films made with sorbitol and $r = -0.72$ for films made with glycerol). This is indicative of a decrease in biodegradability, as the AX polymers were increasingly heterogeneous in the films, which could have impeded the microbial breakdown of the films.

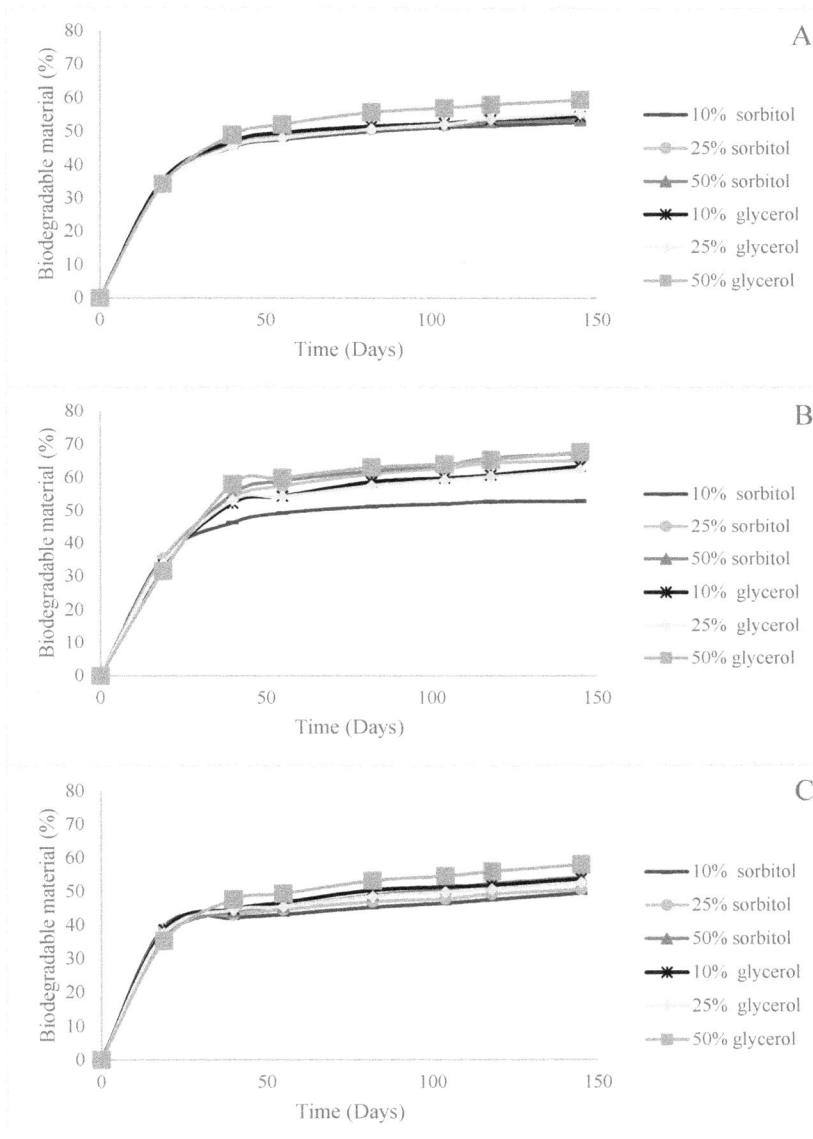

Figure 2. Biodegradability profiles for films made with wheat bran arabinoxylan (**A**), maize bran arabinoxylan (**B**), and dried distillers grain arabinoxylan (**C**).

In addition, the utilization of glycerol instead of sorbitol increased the total biodegradability of the film by about 3%. Furthermore, when comparing the biodegradability of these films, the general trend was that as the plasticizer level increased so did their biodegradability. These trends in the effects of plasticizer type and level on the total amount of biodegradable material in the AX films are most likely due to the loss of order in the polymers of the films. This loss of order creates a material that can be more easily broken down by the microbes in the soil.

Table 4. Carbon contents and total biodegradable material of arabinoxylan films made with arabinoxylan extracted from wheat bran, maize bran, or dried distillers grain.

Film Composition	Carbon Content [a]	Total Biodegradable Material
	(mg)	(%)
WB AX [b] + 10% sorbitol	140.6	52.8 ± 0.59
WB AX + 25% sorbitol	140.6	53.4 ± 1.73
WB AX + 50% sorbitol	141.5	53.4 ± 0.63
WB AX + 10% glycerol	141.4	54.5 ± 0.43
WB AX + 25% glycerol	142.2	55.4 ± 0.34
WB AX + 50% glycerol	135.8	59.5 ± 0.54
MB AX [c] + 10% sorbitol	154.3	53.0 ± 0.40
MB AX + 25% sorbitol	152.1	65.2 ± 0.77
MB AX + 50% sorbitol	145.6	67.3 ± 0.07
MB AX + 10% glycerol	148.3	63.4 ± 1.99
MB AX + 25% glycerol	146.6	62.4 ± 1.25
MB AX + 50% glycerol	134.7	67.7 ± 1.41
DDG AX [d] + 10% sorbitol	161.3	49.4 ± 0.42
DDG AX + 25% sorbitol	159.4	50.5 ± 0.22
DDG AX + 50% sorbitol	157.4	54.1 ± 0.82
DDG AX + 10% glycerol	153.9	53.7 ± 0.29
DDG AX + 25% glycerol	153.6	52.6 ± 1.43
DDG AX + 50% glycerol	143.4	58.0 ± 0.90
LSD ($p < 0.05$)		2.1
LSD ($p < 0.01$)		2.1

[a] Carbon content was only measured a single time for use in the calculation of percent biodegradable material; [b] wheat bran arabinoxylan; [c] maize bran arabinoxylan; [d] dried distillers grain arabinoxylan.

4. Conclusions

Biodegradable packaging material must have a balance between being easily degraded after use and having the proper mechanical properties for the particular application of interest. The materials produced in this study show promise as the basis for biodegradable food packaging materials in the future as demonstrated by their mechanical properties and interactions with water. The general trend in water solubility was as follows for these AX films: WB < MB < DDG. In addition, the DDG AX films were the least biodegradable and the MB AX films were the most biodegradable. These trends in the interactions of these films were related to their physicochemical properties including AX M_w and AX polydispersity index. While there were fewer clear-cut trends between the physicochemical properties of the films and their biodegradability, all films were at least 49% biodegradable. In addition, as the amount of glycerol or sorbitol in the films increased, the films became less hydrophilic but more biodegradable. The combination of these pieces of information can be utilized to tailor biodegradable packaging materials for food products to ensure maximum shelf life. Overall, each type of film tested in this paper can lend itself to various packaging and materials applications. Some of these films may be better suited to plastic wrapping material, while others would be better to use for plastic bags. It would be greatly beneficial to continue researching the properties of AX-based materials to determine how they would behave under all types of environments including anaerobic biodegradation.

Author Contributions: Conceptualization, C.A. and S.S.; Methodology, C.A. and S.S.; Formal Analysis, C.A.; Investigation, C.A. and S.S.; Resources, S.S.; Data Curation, C.A.; Writing-Original Draft Preparation, C.A.; Writing-Review and Editing, C.A. and S.S.; Supervision, S.S.; Project Administration, S.S.; Funding Acquisition, S.S.

Funding: This research received no external funding.

Acknowledgments: The authors would like to thank Kristin Whitney and Chunju Gu for their assistance with this research. In addition, the authors would like to thank the North Dakota State Mill, Agricor, Inc., and Tharaldson Ethanol for donating the starting materials for this research.

Conflicts of Interest: The authors declare no conflict of interest.

References

1. Delcour, J.A.; Hoseney, R.C. (Eds.) Structure of cereals. In *Principles of Cereal Science and Technology*, 3rd ed.; AACC International Inc.: St. Paul, MN, USA, 2010; pp. 1–22.
2. Heikkinen, S.L.; Mikkonen, K.S.; Pirkkalainen, K.; Serimaa, R.; Joly, C.; Tenkanen, M. Specific enzymatic tailoring of wheat arabinoxylan reveals the role of substitution on xylan film properties. *Carbohydr. Polym.* **2013**, *92*, 733–740. [CrossRef] [PubMed]
3. Saulnier, L.; Sado, P.-E.; Branlard, G.; Charmet, G.; Guillon. F. Wheat arabinoxylans: Exploiting variation in amount and composition to develop enhanced varieties. *J. Cereal. Sci.* **2007**, *46*, 261–281. [CrossRef]
4. Maes, C.; Delcour, J.A. Structural characterization of water-extractable and water-unextractable arabinoxylans in wheat bran. *J. Cereal. Sci.* **2002**, *35*, 315–326. [CrossRef]
5. Zhang, Y.; Pitkänen, L.; Douglade, J.; Tenkanen, M.; Remond, C.; Joly, C. Wheat bran arabinoxylans: Chemical structure and film properties of three isolated fractions. *Carbohydr. Polym.* **2011**, *86*, 852–859. [CrossRef]
6. Kiszonas, A.M.; Fuerst, E.P.; Morris, C.F. Wheat arabinoxylan structure provides insight into function. *Cereal Biomacromol.* **2013**, *90*, 387–395. [CrossRef]
7. Reis, S.F.; Coelho, E.; Coimbra, M.A.; Abu-Ghannam, N. Improved efficiency of brewer's spent grain arabinoxylans by ultrasound-assisted extraction. *Ultrason. Sonochem.* **2015**, *24*, 155–164. [CrossRef]
8. BeMiller, J.N. Monosaccharides. In *Carbohydrate Chemistry for Food Scientists*, 2nd ed.; BeMiller, J.N., Ed.; AACC International Inc.: St. Paul, MN, USA, 2007; pp. 1–24. [CrossRef]
9. Zhang, Z.; Smith, C.; Li, W. Extraction and modification technology of arabinoxylans from cereal by-prroducts: A critical review. *Food Res. Int.* **2014**, *65*, 423–436. [CrossRef]
10. Anson, M.N.; Hemery, Y.M.; Bast, A.; Haenen, G.R. Optimizing the bioactive potential of wheat bran by processing. *Food Funct.* **2012**, *3*, 362–375. [CrossRef]
11. Delcour, J.A.; Hoseney, R.C. (Eds.) Minor constituents. In *Principles of Cereal Science and Technology*, 3rd ed.; AACC International Inc.: St. Paul, MN, USA, 2010; pp. 71–86. [CrossRef]
12. Saeed, F.; Pasha, I.; Anjum, F.M.; Sultan, M.T. Arabinoxylans and arabinogalactans: A comprehensive treatise. *Food Sci. Nutr.* **2011**, *51*, 467–476. [CrossRef]
13. Aguedo, M.; Fougnies, C.; Dermience, M.; Richel, A. Extraction by three processes of arabinoxylans from wheat bran and characterization of the fractions obtained. *Carbohydr. Polym.* **2014**, *105*, 317–324. [CrossRef]
14. Phan The, D.; Debeaufort, F.; Peroval, C.; Despre, D.; Courthaudon, J.L.; Voilley, A. Arabinoxylan lipid-based edible films and coatings. *J. Agric. Food Chem.* **2002**, *50*, 2423–2428. [CrossRef] [PubMed]
15. Zhang, P.; Whistler, R.L. Mechanical properties and water vapor permeability of thin film from corn hull arabinoxylan. *J. Appl. Polym. Sci.* **2004**, *93*, 2896–2902. [CrossRef]
16. Casariego, A.; Souza, B.W.S.; Cerqueira, M.A.; Teixeira, J.A.; Cruz, L.; Díaz, R.; Vicente, A.A. Chitosan/clay films' properties as affected by biopolymer and clay micro/nanoparticles' concentrations. *Food Hydrocoll.* **2009**, *23*, 1895–1902. [CrossRef]
17. Tharanathan, R.N. Biodegradable films and composite coatings: Past, present and future. *Trends Food Sci. Technol.* **2003**, *14*, 71–78. [CrossRef]
18. Anderson, C.; Simsek, S. Mechanical profiles and topographical properties of films made from alkaline extracted arabinoxylans from wheat bran, maize bran, or dried distillers grain. *Food Hydrocoll.* **2018**. [CrossRef]
19. Mendis, M.; Simsek, S. Production of structurally diverse wheat arabinoxylan hydrolyzates using combinations of xylanase and arabinofuranosidase. *Carbohydr. Polym.* **2015**, *132*, 452–459. [CrossRef] [PubMed]
20. AACC International. Method 08-01.01. Ash-Basic Method. In *Approved Methods of Analysis*, 11th ed.; AACC International, Inc.: St. Paul, MN, USA, 1999; Volume 1234.
21. AACC International. Method 44-15.02. Moisture-Air Oven Methods. In *Approved Methods of Analysis*, 11th ed.; AACC International Inc.: St. Paul, MN, USA, 1999. [CrossRef]
22. AACC International. Method 46-30.01. Crude Protein-Combustion Method. In *Approved Methods of Analysis*, 11th ed.; AACC International Inc.: St. Paul, MN, USA, 1999. [CrossRef]
23. Blakeney, A.B.; Harris, P.J.; Henry, R.J.; Stone, B.A. A simple and rapid preparation of alditol acetates. *Carbohdr. Res.* **1983**, *113*, 291–299. [CrossRef]
24. Wyatt Technology. *Astra Software*, Version 6.0.5; Wyatt Technology: Santa Barbra, CA, USA, 2016.

25. Dervilly, G.; Saulnier, L.; Roger, P.; Thibault, J.-F. Isolation of homogeneous fractions from wheat water-soluble arabionxylans. *J. Agric. Food Chem.* **2000**, *48*, 270–278. [CrossRef]

26. Bruker BioSpin Corporation. *TopSpin*, Bruker BioSpin Corporation: Billerica, MA, USA, 2015.

27. Garcia, M.; Pinotti, A.; Martino, M.; Zaritzky, N. Characterization of composite hydrocolloid films. *Carbohydr. Polym.* **2004**, *56*, 339–345. [CrossRef]

28. ASTM International. *Standard Test Method for Measurement of the Surface Tension of Solid Coatings, Substrates and Pigments Using Contact Angle Measurements*; ASTM Standard D7490-13; ASTM International: West Conshohocken, PA, USA, 2013. [CrossRef]

29. ASTM International. *Standard Test Methods for Water Vapor Transmission of Materials*; ASTM Standard E96/E96M-15; ASTM International: West Conshohocken, PA, USA, 2015. [CrossRef]

30. Colussi, R.; Pinto, V.Z.; El Halal, S.L.M.; Biduski, B.; Prietto, L.; Castibos, D.D.; Zawareze, E.R.; Dias, A.R.G. Acetylated rice starches films with different levels of amylose: Mechanical, barrier and thermal properties and biodegradability. *Food Chem.* **2017**, *221*, 1614–1620. [CrossRef]

31. ASTM International. *Standard Test Method for Determining Aerobic Biodegradation of Plastic Materials in Soil*; ASTM Standard D5988-12; ASTM International: West Conshohocken, PA, USA, 2012. [CrossRef]

32. SAS Institute Inc. *SAS Software*, SAS Institute Inc.: Cary, NC, USA, 2011.

33. Mandalari, G.; Faulds, C.B.; Sancho, A.I.; Saija, A.; Bisignano, G.; LoCurto, R.; Waldron, K.W. Fractionation and characterisation of arabinoxylans from brewers' spent grain and wheat bran. *J. Cereal Sci.* **2005**, *42*, 205–212. [CrossRef]

34. Müller, C.M.O.; Yamashita, F.; Laurindo, J.B. Evaluation of the effects of glycerol and sorbitol concentration and water activity on the water barrier properties of cassava starch films through a solubility approach. *Carbohydr. Polym.* **2008**, *72*, 82–87. [CrossRef]

35. Antoniou, J.; Liu, F.; Majeed, H.; Qazi, H.J.; Zhong, F. Physicochemical and thermomechanical characterization of tara gum edible films: Effect of polyols as plasticizers. *Carbohydr. Polym.* **2014**, *111*, 359–365. [CrossRef] [PubMed]

36. Nazan Turhan, K.; Şahbaz, F. Water vapor permeability, tensile properties and solubility of methylcellulose-based edible films. *J. Food Eng.* **2004**, *61*, 459–466. [CrossRef]

37. Casariego, A.; Souza, B.W.S.; Vicente, A.A.; Teixeira, J.A.; Cruz, L.; Díaz, R. Chitosan coating surface properties as affected by plasticizer, surfactant and polymer concentrations in relation to the surface properties of tomato and carrot. *Food Hydrocoll.* **2008**, *22*, 1452–1459. [CrossRef]

38. Lawoko, M.; Nutt, A.; Henriksson, H.; Gellerstedt, G.; Henroksson, G. Hemicellulase activity of aerobic fungal cellulases. *J. Appl. Microbiol.* **1999**, *87*, 366–370. [CrossRef]

39. Shallom, D.; Shoham, Y. Microbial hemicellulases. *Curr. Opin. Microbiol.* **2003**, *6*, 219–228. [CrossRef]

40. Gunnarsson, S.; Marstorp, H.; Dahlin, A.S.; Witter, E. Influence of non-cellulose structural carbohydrate composition on plant material decomposition in soil. *Biol. Fertil. Soils* **2008**, *45*, 27–36. [CrossRef]

Article

Effects of Drying Methods and Ash Contents on Heat-Induced Gelation of Porcine Plasma Protein Powder

Chengli Hou, Wenting Wang, Xuan Song, Liguo Wu and Dequan Zhang *

Institute of Food Science and Technology, Chinese Academy of Agricultural Sciences/Key Laboratory of Agro-Products Processing, Ministry of Agriculture and Rural Affairs, Beijing 100193, China; houchengli@163.com (C.H.); wangwentingbiome@163.com (W W.); songxuan120@163.com (X.S.); liguowu911@163.com (L.W.)

* Correspondence: dequan_zhang0118@126.com or zhangdequan@caas.cn; Tel.: +86-10-62818740; Fax: +86-10-62818740

Received: 5 April 2019; Accepted: 22 April 2019; Published: 25 April 2019

Abstract: Porcine blood plasma is a rich source of proteins with high nutritional and functional properties, which can be used as a food ingredient. The plasma is usually processed into powders in applications. In the present study, the effects of drying methods and ash contents on heat-induced gelation of plasma protein powder were investigated. The drying methods had a significant impact on the gel properties of the plasma powder heat-induced gels. The hardness and elasticity of the gels by freeze-dried and spray-dried plasma powders were lower than that of the liquid plasma ($p < 0.05$). The microstructures of dehydrated plasma were denser and the holes were smaller. The secondary structure of the gels from the spray-dried plasma protein powders exhibited more α-helixes and less β-turns than that from the freeze-dried powder and liquid plasma. The thermostability of dehydrated plasma powder was found to have decreased compared to the liquid plasma. Compared with the gels obtained from the high ash content plasma protein powders, the gel from the 6% ash content plasma powder had the highest water-holding capacity and had the lowest hardness and elasticity. However, the secondary structure and microstructures of the heat-induced gels were not affected by the ash contents in the plasma powders. These findings show that the gel properties of plasma protein powder can be finely affected by drying methods and ash contents.

Keywords: blood plasma protein powder; heat-induced gelation; drying method; ash content; texture

---

1. Introduction

Blood is one of the main coproducts from slaughtered animals. The yield of animal blood has reached approximately 10 million tons per year all around the world [1,2]. Animal blood is typically discarded as waste, which causes environmental pollution [3]. Plasma is the product of anticoagulant blood after centrifugation and the removal of blood cells. Blood plasma is a rich source of proteins with high nutritional and functional quality. Most plasma proteins are used as ingredients, mainly as binders, emulsifiers, fat replacers, polyphosphate replacements, and meat curing agents in the food industry [3–7]. Gel-forming ability upon heating is considered to be the most interesting attribute of plasma. Previous studies have reported that the heat-induced gelation of plasma was affected by pH and cysteine [8–12].

Plasma contains a complex mixture of proteins. The typical composition is 50–60% albumin, 40–50% globulins, and 1–3% fibrinogen [10]. Liquid plasma is dried to a powder for better storage and transportation. The protein powder processing eliminates many disadvantageous factors, such as perishability and difficulty to store and transport [13,14]. Spray-drying and freeze-drying are commonly used as a dehydration technique for making protein powder products. Spray-drying has many merits, such as simple operation, drying quickly, low cost, and being suitable for continuous mass production. However, spray-drying may lead to protein denaturation and conformation changes due to a higher temperature, which will probably affect the plasma protein function [15,16]. Freeze-drying can sublimate the moisture in the material directly at a low temperature by doing less harm to the structure of the protein. However, the main disadvantage of freeze-drying is the high cost. A previous study showed that the changes in temperature, moisture, and salt ions during drying could affect the quality of the protein products [17].

Ash is an important factor that can affect the quality of plasma protein powder. It mainly comes from mineral ions in blood and exogenous anticoagulants. According to previous research, the content of ash in plasma protein powder is usually up to 14% [18]. The main ingredients of ash include sodium, magnesium, and calcium, most of which is sodium [18]. Plasma protein powder with low ash content can be produced by concentrating and removing the salt with ultrafiltration and nanofiltration. Sodium chloride (NaCl) at a concentration of 1–3% is needed to facilitate protein solubilization, resulting in the gel [19]. However, the effect of ash content on the functional properties of plasma protein gel is not clear.

The objective of this study was to determine the effects of two different dehydration methods on plasma protein functionality. Additionally, on this basis, the effect of the ash content of freeze-dried plasma on the gelation properties was studied. The results can provide data support for product development and application of plasma protein powder as a food additive.

2. Materials and Methods

2.1. Materials

Fresh porcine blood was obtained from a local slaughterhouse in Beijing, China. Sodium citrate (0.345% w/v final concentration) was added to prevent coagulation. Plasma was separated by centrifugation at 2437× *g* for 8 min at 4 °C (Himac CR22 GII, Hitachi, Ltd., Tokyo, Japan).

2.2. Preparation of Plasma Powders

Freeze-dried plasma powder was obtained after 72 h of lyophilization, under the following conditions: cold trap temperature, −60 °C; material temperature, −40 °C; vacuum, 1 Pa (LGJ-25, Four-Ring Science Instrument Plant Beijing Co., Ltd, Beijing, China). Spray-dried plasma powder (nitrogen solubility index 98.92%) was prepared using a laboratory scale spray dryer (SD-Basic, Labplant, UK) under the following conditions: inlet temperature, 150 °C; feeding speed, 0.178 mL/s. The salt ions of plasma were removed using a multi-stage membrane separation experimental machine (DMJ60-2, Jinan Bona Biological Technology Co., Ltd, Shandong, China), and freeze-dried to prepare the plasma powder with the lowest ash content. The cut-off molecular weight of the coil ultrafiltration membrane components was 10,000 Da, and the selected molecular weight of the coil nanofiltration membrane was 150 Da. The membrane's working pressure was 0.6 MPa and the pH of the plasma was 9. The ash content of the desalted and the un-desalted freeze-dried plasma protein powder were determined according to the Chinese standard GB 5009.4 [20]. Different proportions of the freeze-dried plasma protein powder, without the treatment of ultrafiltration, were mixed with the freeze-dried desalted plasma powder to prepare the plasma powder with different ash contents (calculated values: 7%, 9%, 12%, 15%, and 18%, respectively; measured values with the same unit protein content: 6.17%, 8.92%, 11.92%, 15.24%, and 18.82%, respectively).

2.3. Water-Holding Capacity and Texture Analyses

Plasma protein powders were dissolved in ultrapure water. The protein concentrations of liquid plasma and plasma powder solutions were adjusted to 60 mg/mL, and the pH of these solutions was adjusted to 9. Samples were heated at 80 °C for 45 min to form gels. The gels were immediately cooled to room temperature and stored in the refrigerator at 4 °C overnight to age for further analysis [16]. The water holding capacity (WHC) of the gels was calculated via a centrifugal method [21]. The pieces of the gel after weighing were centrifuged at 1000× g for 10 min at 4 °C. WHC was calculated as the percentage of water retained based on the water content in the gels prior to centrifugation. Three replicates were measured for each sample.

Textures (hardness and elasticity) were analyzed by the texture profile analysis (TPA) test using a texture analyzer (TA-XT2i/5, Stable Micro Systems, Godalming, UK) with a cylindrical probe (P/0.5R) according to the method of Li et al [22]. The parameters were as follows: pre-test speed, 1.0 mm/s; test speed, 0.5 mm/s; withdrawal speed, 1.0 mm/s; depth of probe penetration, 5 mm; minimum trigger force, 5 g; and data acquisition rate, 200 points/s. For each gel, the texture was measured in triplicate.

2.4. Microstructure

The microstructures of the gels were investigated by scanning electron microscopy (SEM). The gels formed from liquid and dehydrated plasma were fixed in 3% glutaraldehyde for 48 h at 4 °C, re-fixed in osmic acid for 2 h, and washed three times with phosphate-buffered saline. Then, they were dehydrated using an ethanol series (50%, 70%, 80%, 90%, and 100% (v/v ethanol, successively)). Freeze-drying and sputter-coating were performed according to the procedure of Han et al [23]. Samples were dried using carbon dioxide critical point drying, and coated with Au in a vacuum ion sputtering system. These specimens were observed in a Hitachi SU8010 SEM (Hitachi Ltd., Tokyo, Japan), operating at a voltage of 15 kV [24].

2.5. Fourier Transform Infrared Measurements

The gels of plasma powders and liquid plasma were prepared in an 80 °C water bath for 45 min and were measured using a Fourier transform infrared spectrometer detector (Tensor 27, Bruker, Germany). The resolution was 4 cm^{-1}, the scanning range was 4000–600 cm^{-1}, and the signal was cumulatively scanned 64 times. Spectral OPUS 7.0 software was used for background subtraction and CO_2 atmosphere compensation. Peakfit 4.2 was used for baseline correction, a second derivative peak fitting of gauss points of the amide Iband (1600–1700 cm^{--}), and the estimated position and number of the stack peak of the amide I band.

2.6. Differential Scanning Calorimetry

The level of denaturation for plasma proteins was studied using differential scanning calorimetry (DSC). One hundred microliters of liquid and dehydrated plasma solutions with the same protein concentration (60 mg/mL) were placed in DSC pans, hermetically sealed, and subsequently analyzed using a DSC Q200 (TA Instrument, New Castle, DE, USA). A pan that contained 100 µL of distilled water was used as a control. All pans were heated from 25 to 105 °C at 3 °C/min. The thermal denaturation point (T_d, °C), which was the minimal heat flows in the DSC thermogram, and the enthalpy of denaturation (ΔH, J/g) determined by the integration of the area belonging to the changes in heat flow, as a function of the temperature, was calculated from the thermogram [25].

2.7. Statistical Analysis

All heat-induced gels were carried out in triplicate and each sample was measured in triplicate. Analysis of variance (ANOVA) was performed using SPSS 22 for Windows (SPSS Inc., Chicago, IL, USA). The differences of means were evaluated by the Duncan test ($p \leq 0.05$).

3. Results and Discussion

3.1. Effect of Drying Methods on Heat-Induced Gelation of Plasma Proteins

Previous studies reported that the gel properties of plasma protein can be finely adjusted by pH [10,11,16]. The hardness of heat-induced gels can be increased by increasing the pH levels [8]. In the present study, pH 9 was selected, as the gels have good gel properties in this condition.

WHC is one of the most important functional properties of heat-induced protein gels [26]. In the present study, the WHC was not significantly different among the three groups ($p > 0.05$) (Figure 1A). The result was in agreement with Parés et al. [16], who reported that the WHC was not different between gels obtained from liquid plasma and spray-dried plasma, at any given pH (4.5, 5.5, 6, and 7.4). However, the result was not in agreement with Gong et al. [27] working with peanut protein isolate, which showed that the WHC of freeze-dried peanut protein isolate were significantly higher than those of the spray-dried one. The high temperature of spray-drying can affect the structure and properties of the protein [28,29]. This could be due to the differences in the protein and drying parameters.

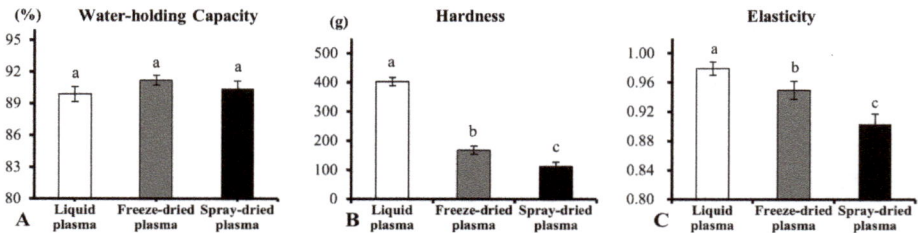

Figure 1. The effect of drying methods on water-holding capacity (WHC) and texture of heat-induced porcine plasma protein gels. (**A**) WHC of the gels, (**B**) Hardness of the gels, (**C**) Elasticity of the gels. Different letters indicate significant differences ($p < 0.05$).

The hardness and elasticity of heat-induced gels significantly decreased in the freeze-dried and the spray-dried plasma powders ($p < 0.05$) (Figure 1B,C). Besides this, the hardness and elasticity of the gels from the spray-dried plasma were less than that of the freeze-dried plasma. Due to the ion concentration changes, the degeneration of plasma protein occurs during dehydration of the liquid plasma [16]. The spray-drying may cause more plasma proteins to be denatured on heating compared to freeze-drying [29,30]. The denatured proteins effect the aggregation of proteins [16,30], which may lead to a reduction in gel hardness and elasticity.

The microstructures of the heat-induced liquid plasma and dehydrated plasma powder gels are shown in Figure 2. The gels exhibit a clearly ordered porous structure, and slight differences were observed among different samples. The pores of gels from liquid plasma were slightly larger than those of dehydrated plasma powders (Figure 2B,C). Wang et al. [11] reported that fine-stranded gels were formed when the pH was higher than 6.0, but a disordered and particulate gel network with several large pores was formed at a low pH, i.e., 5.5. The present result showed that fine-stranded gels were formed for liquid plasma at pH 9, which is consistent with the previous study. Parés et al. [16] showed that no notable differences in the microscopic structure of gels from liquid and spray-dried plasma were observed. In the present study, there were only slight differences between the treatments. The microstructure of the freeze-dried plasma gel was more compact than that of the liquid plasma gel. These structural modifications could explain no differences in WHC, but the hardness and elasticity of dehydrated plasma are lower than those of liquid plasma.

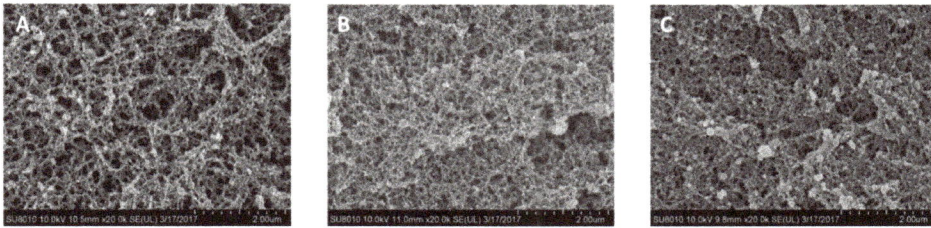

Figure 2. Scanning electron micrograph (magnification, ×20,000) of heat-induced porcine plasma protein gels from different drying methods. (**A**) Gels from liquid plasma, (**B**) Gels from freeze-dried plasma, (**C**) Gels from spray-dried plasma.

The secondary structure of the gels from liquid plasma and spray-dried and freeze-dried plasma powders are shown in Figure 3. The secondary structure of gels from the liquid plasma and freeze-dried plasma powders were similar. The main secondary structure was β-sheet, followed by β-turn; there were fewer random coils and α-helixes The secondary structure of the gels from spray-dried plasma protein powders exhibited a different composition. The main secondary structure was β-sheet, followed by β-turn and α-helix, and there were fewer random coils. A previous study has shown that the spray-dried peanut protein isolate had a relatively more unfolded or flexible structure than the freeze-dried peanut protein isolate [27]. In the present study, the result also showed that spray-drying affected the structure of protein gels. The reason is that the thermal denaturation process significantly affected the protein's secondary structure [31].

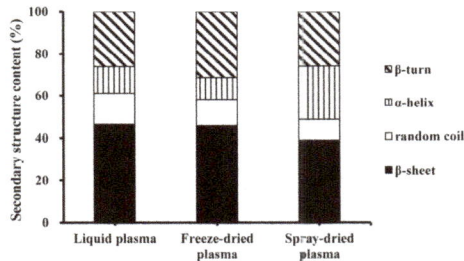

Figure 3. The effect of drying methods on the secondary structure of heat-induced porcine plasma protein gels.

The DSC curves of the liquid plasma and spray-dried and freeze-dried plasma powders are shown in Figure 4. The T_d of liquid plasma was significantly higher than that of the dehydrated plasma ($p < 0.05$), indicating that the thermal stability of the liquid plasma was better. The thermal denaturation of protein is closely related to the change in its spatial conformation. The thermal stability of plasma proteins was changed during drying, which led to the different thermal denaturation states. The ΔH of the liquid plasma and spray-dried plasma were higher than that of the freeze-dried plasma. These results were not in agreement with the study of Parés et al. [16], who reported that the differences of the peak temperature and enthalpy calculated for the liquid plasma and spray-dried plasma were not significant at the same pH (4.5, 5.5, 6.0, and 7.4).

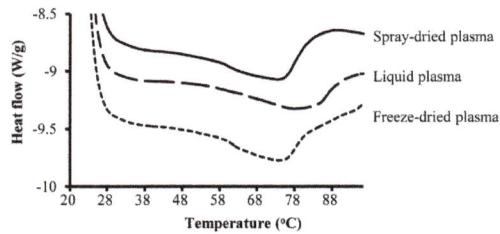

	Liquid plasma	Freeze-dried plasma	Spray-dried plasma
T_d (°C)	78.95 ± 0.47 [a]	74.46 ± 0.03 [b]	74.47 ± 0.15 [b]
ΔH (J/g)	3.05 ± 0.30 [a]	1.78 ± 0.22 [b]	3.01 ± 0.13 [a]

Figure 4. The effect of drying methods on the thermal denaturating temperature of plasma protein. Different letters indicate significant differences ($p < 0.05$).

3.2. Effect of Ash Contents on Heat-Induced Gelation of Plasma Protein

In the current study, the freeze-dried plasma powder showed better gel properties (higher hardness and elasticity) than the spray-dried plasma powder. On this basis, the effect of the ash content of freeze-dried plasma on the gelation properties was studied. The content of the ash is an important indicator of plasma protein powder products. The main component of ash is sodium, then potassium, and calcium. Research shows that sodium chloride affects the gel properties of the protein [32,33]. In Figure 5A, the WHC of heat-induced gels decreased with the increasing ash contents of freeze-dried plasma protein powders. The WHC for the samples with 6% and 9% ash content was significantly higher than that for the samples with 12%, 15%, and 19% ash content ($p < 0.05$). No significant difference was found as ash content increased from 12% to 19%. In the present study, we found that the gel from the 6% plasma protein powder had a soft texture and high viscosity (Figure 5B,C). The ash content of plasma powder significantly influenced the hardness and elasticity of heat-induced gels ($p < 0.05$). The hardness of heat-induced gels increased first and then decreased with the increasing ash content. The gel of the sample with 6% ash content has the lowest hardness and elasticity. The gel of the sample with 15% ash content has the highest hardness and elasticity. The elasticity values were not different between 9%, 12%, and 15% ash content samples. Those results indicated that the texture of gels with low ash content were worse compared to high ash content plasma protein powder. However, the gels with low ash content had a good WHC. This result could be in agreement with that obtained by Meng et al. [34], emphasizing that the WHC decreased with the ion concentration, increasing if the ion concentration was larger than 0.3 mol/L.

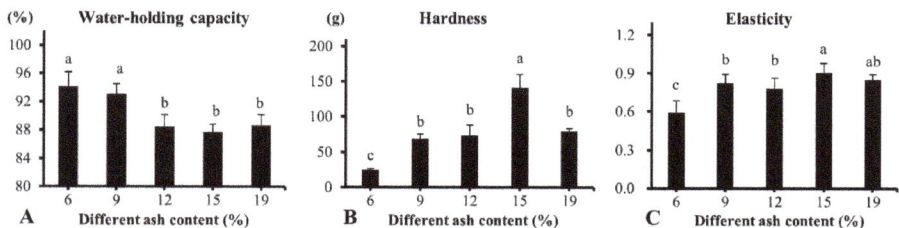

Figure 5. Effect of different ash contents on the water-holding capacity (WHC) and texture of heat-induced porcine plasma protein gels. (**A**) WHC of the gels, (**B**) Hardness of the gels, (**C**) Elasticity of the gels. Different letters indicate significant differences ($p < 0.05$). The corresponding unit ash contents of 6%, 9%, 12%, 15%, and 19% were 6.17%, 8.92%, 11.92%, 8.92%, and 18.82%, respectively, while adjusting the concentration of each protein to be consistent.

During the formation of the gel, the sodium neutralizes the charge on the surface of the protein, leading to the attraction between protein molecules enhancing, and the molecules rapidly aggregating to form a hard gel. When the ash content increases to a certain extent, it is difficult to form gel because of the high concentration of salt-stabilized protein molecular conformations. A previous study showed that high concentrations of NaCl decreased the water-holding capacity of egg-white gels [35], and the present study showed the same result. The reason for this can be attributed to the unstable water molecules trapped in large cavities in the protein gel network [36], and the high solid content existing in the plasma proteins.

The microstructures of the heat-induced gels for different ash content plasma protein powders are shown in Figure 6. The results showed that ordered and three-dimensional network gels were formed. The micrographs did not distinguish between different ash content plasma powders. Therefore, if we just focus on the microstructure of plasma gels, low ash content had little effect on the gels' microstructure.

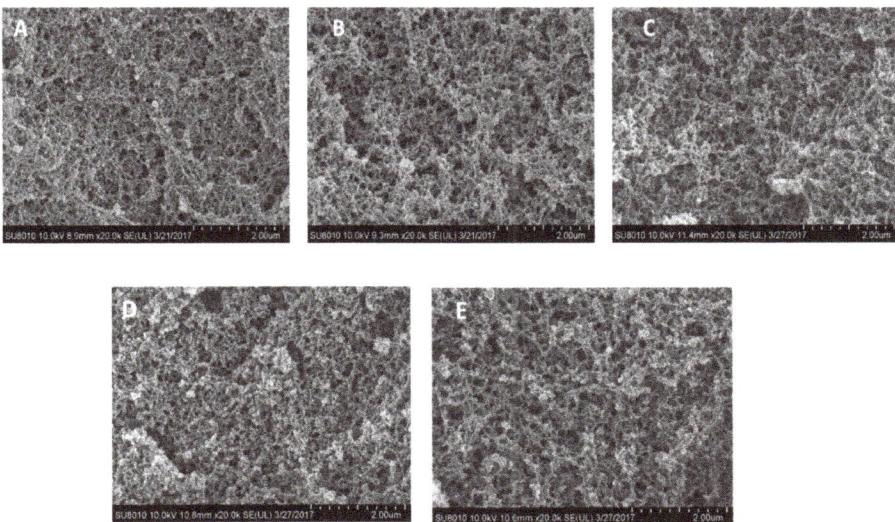

Figure 6. Scanning electron micrographs (magnification ×20,000) of heat-induced porcine plasma protein gels with different ash contents. (**A**) Gels from plasma powder with 6% ash content, (**B**) Gels from plasma powder with 6% ash content, (**C**) Gels from plasma powder with 12% ash content, (**D**) Gels from plasma powder with 15% ash content, (**E**) Gels from plasma powder with 19% ash content,. The ash of A–E were 6%, 9%, 12%, 15%, and 19%, and their corresponding unit ash content was 6.17%, 8.92%, 11.92%, 15.24%, and 18.82%, respectively, while adjusting the concentration of each protein to be consistent.

The secondary structures of plasma protein powders with different ash contents are shown in Figure 7. The secondary structures of gels from different ash content plasma protein powders were similar. Furthermore, the main secondary structure of the heat-induced gel was β-sheet, followed by β-turn, and there were fewer random coils and α-helixes. A previous study reported that some physical and chemical conditions–pH, ion concentration, sugar content, and metal content of protein solution–affected the protein's secondary structure [37]. However, the present result showed that the influence of the ash content (6–19%) on the composition and content of the secondary structure was insignificant.

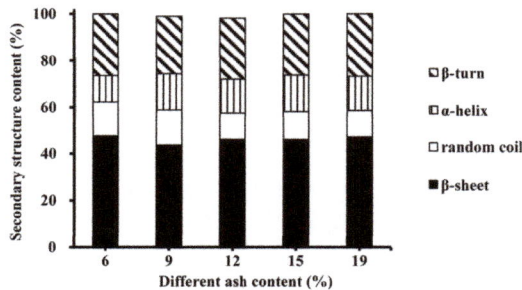

Figure 7. Effect of different ash contents on the secondary structure of heat-induced porcine plasma protein gels.

4. Conclusions

The gels from the dehydrated plasma powders exhibited lower hardness and elasticity than that from the liquid plasma. A possible cause is that dehydrated plasma had lower thermostability and formed a gel with dense microstructures. The gel from the spray-dried plasma powder exhibited lower hardness and elasticity than that from the freeze-dried plasma powder. The secondary structure of the gels from the spray-dried plasma protein powders exhibited more α-helixes and less β-turns than that from the freeze-dried plasma protein and liquid plasma. Compared with the gels of high ash content plasma protein powders, the gel from the 6% ash content plasma powder had the highest water-holding capacity and had the lowest hardness and elasticity. However, the secondary structure and microstructures of the heat-induced gels were not affected by the ash contents of plasma powders. Therefore, drying methods and the ash contents of plasma protein powders affect the quality of heat-induced gel properties. Further studies on food model systems are necessary to confirm the results obtained in the present study.

Author Contributions: C.H. and D.Z. conceived and designed the experiments; W.W., C.H., and X.S. performed the experiments. C.H., W.W., and D.Z. analyzed the data and wrote the paper. X.S. and L.W. helped perform the analysis with constructive discussions.

Funding: This research was financially supported by the Fundamental Research Funds for the Central Non-Profit Scientific Institution (No. S2016JC11), the Modern Agricultural Talent Support Program-Outstanding Talents and Innovative Team of Agricultural Scientific Research (2016-2020), and the National High-level personnel of special support program and National Agricultural Science and Technology Innovation Program in China.

Acknowledgments: We wish to thank Elena Saguer (University of Girona, Spain) for her help and advice during the course of this work. We would also like to thank Beijing Ershang Group Dahongmen Meat Food Co., Ltd. for their help with blood sampling.

Conflicts of Interest: The authors declare no conflict of interest.

References

1. Liu, Y.D.; Wu, H.L.; Zhang, J.; Wan, D.J.; Zhao, J.H. Study of porcine blood decoloration technique. *Adv. Mater. Res.* **2012**, *518–523*, 3980–3983. [CrossRef]
2. Mielink, J.; Slinde, E. Sausage color measured by integrating sphere reflectance spectrophotometry when whole blood or blood cured by nitrite is added to sausages. *J. Food Sci.* **2010**, *48*, 1723–1725. [CrossRef]
3. Ofori, J.A.; Hsieh, Y.H.P. *The Use of Blood and Derived Products as Food Additives*; El-Samragy, Y., Ed.; InTech: Rijeka, Croatia, 2012.
4. Fowler, M.R.; Park, J.W. Effect of salmon plasma protein on Pacific whiting surimi gelation under various ohmic heating conditions. *LWT Food Sci. Technol.* **2015**, *61*, 309–315. [CrossRef]
5. Romero de Ávila, M.D.; Ordóñez, J.A.; Escudero, R.; Cambero, M.I. The suitability of plasma powder for cold-set binding of pork and restructured dry ham. *Meat Sci.* **2014**, *98*, 709–717. [CrossRef] [PubMed]

6. Ni, N.; Wang, Z.; Wang, L.; He, F.; Liu, J.; Gao, Y.; Zhang, D. Reduction of sodium chloride levels in emulsified lamb sausages: The effect of lamb plasma protein on the gel properties, sensory characteristics, and microstructure. *Food Sci. Biotechnol.* **2014**, *23*, 1137–1143. [CrossRef]

7. Hurtado, S.; Saguer, E.; Toldrà, M.; Parés, D.; Carretero, C. Porcine plasma as polyphosphate and caseinate replacer in frankfurters. *Meat Sci.* **2012**, *90*, 624–628. [CrossRef] [PubMed]

8. Ni, N.; Wang, Z.; Chen, L.; Xu, W.; Pan, H.; Gao, Y.; Zhang, D. Effect of pH on the gel properties of lamb plasma protein during heat-induced gelation. *Mod. Food Sci. Technol.* **2015**, *31*, 160–166.

9. Saguer, E.; Alvarez, P.; Fort, N.; Espigulé, E.; Parés, D.; Toldrà, M.; Carretero, C. Heat-induced gelation mechanism of blood plasma modulated by cysteine. *J. Food Sci.* **2015**, *80*, 515–521. [CrossRef]

10. Dàvila, E.; Parés, D.; Cuvelier, G.; Relkin, P. Heat-induced gelation of porcine blood plasma proteins as affected by pH. *Meat Sci.* **2007**, *76*, 216–225. [CrossRef] [PubMed]

11. Wang, P.; Xu, X.; Huang, M.; Huang, M.; Zhou, G. Effect of pH on heat-induced gelation of duck blood plasma protein. *Food Hydrocoll.* **2014**, *35*, 324–331. [CrossRef]

12. Eduard, D.; Dolors, P.; Howell, N.K. Fourier transform raman spectroscopy study of heat-induced gelation of plasma proteins as influenced by pH. *J. Agric. Food Chem.* **2006**, *54*, 7890–7897.

13. Huang, Q.; Ma, C.J.; Zhou, J.; Ma, M.H. Effects of drying methods and physico-chemical factors on functional properties of quail egg-white protein powder. *Food Sci.* **2008**, *29*, 299–302.

14. Qiu, C.Y.; Sun, W.Z.; Cui, C.; Zhao, M.M. Effects of drying methods on characteristics of deamidated wheat gluten. *J. South China Univ. Technol.* **2014**, *42*, 129–135.

15. Linarès, E.; Larré, C.; Popineau, Y. Freeze- or spray-dried gluten hydrolysates. 1. Biochemical and emulsifying properties as a function of drying process. *J. Food Eng.* **2001**, *48*, 127–135. [CrossRef]

16. Parés, D.; Saguer, E.; Saurina, J.; Sunol, J.J.; Carretero, A.C. Functional properties of heat induced gels from liquid and spray-dried porcine blood plasma as influenced by pH. *J. Food Sci.* **1998**, *63*, 958–961. [CrossRef]

17. Kinsella, J.E.; Srinivasan, D. Nutritional, chemical, and physical criteria affecting the use and acceptability of proteins in foods. In *Criteria of Food Acceptance*; Solms, J., Hall, R.L., Eds.; Foster Verlag: Zürich, Switzerland, 1981.

18. Wang, W.; Hou, C.; Song, X.; Li, Z.; Wu, L.; Boga, L.A.I.; Zhu, J.; Zhang, D. Comparative and analysis of the quality of plasma protein powder. *Food Sci. Technol. Int.* **2017**, *42*, 119–125. [CrossRef] [PubMed]

19. Kim, Y.S.; Park, J.W. Negative roles of salt in gelation properties of fish protein isolate. *J. Food Sci.* **2008**, *73*, C585–C588. [CrossRef]

20. National Health and Family Planning Commission of China. *Chinese Standard GB 5009.4-2016. Determination of Ash in Foods*; China Standards Press of China: Beijing, China, 2016.

21. Sun, J.; Wu, Z.; Xu, X.; Li, P. Effect of peanut protein isolate on functional properties of chicken salt-soluble proteins from breast and thigh muscles during heat-induced gelation. *Meat Sci.* **2012**, *91*, 88–92. [CrossRef]

22. Li, Y.; Kong, B.; Xia, X.; Liu, Q.; Diao, X. Structural changes of the myofibrillar proteins in common carp (*Cyprinus carpio*) muscle exposed to a hydroxyl radical-generating system. *Process Biochem.* **2013**, *48*, 863–870. [CrossRef]

23. Han, M.; Zhang, Y.; Fei, Y.; Xu, X.; Zhou, G. Effect of microbial transglutaminase on NMR relaxometry and microstructure of pork myofibrillar protein gel. *Eur. Food Res. Technol.* **2009**, *228*, 665–670. [CrossRef]

24. Wu, M.; Xiong, Y.L.; Chen, J. Rheology and microstructure of myofibrillar protein-plant lipid composite gels: Effect of emulsion droplet size and membrane type. *J. Food Eng.* **2011**, *106*, 318–324. [CrossRef]

25. Kamimura, J.A.; Santos, E.H.; Hill, L.E.; Gomes, C.L. Antimicrobial and antioxidant activities of carvacrol microencapsulated in hydroxypropyl-beta-cyclodextrin. *LWT Food Sci. Technol.* **2014**, *57*, 701–709. [CrossRef]

26. Puolanne, E.; Halonen, M. Theoretical aspects of water-holding in meat. *Meat Sci.* **2010**, *86*, 151–165. [CrossRef] [PubMed]

27. Gong, K.; Shi, A.; Liu, H.; Liu, L.; Hu, H.; Adhikari, B.; Wang, Q. Emulsifying properties and structure changes of spray and freeze-dried peanut protein isolate. *J. Food Eng.* **2016**, *170*, 33–40. [CrossRef]

28. Samsalee, N.; Sothornvit, R. Effect of natural cross-linkers and drying methods on physicochemical and thermal properties of dried porcine plasma protein. *Food Biosci.* **2017**, *19*, 26–33. [CrossRef]

29. Anandharamakrishnan, C.; Rielly, C.D.; Stapley, A.G.F. Effects of process variables on the denaturation of whey proteins during spray drying. *Dry. Technol.* **2007**, *25*, 799–807. [CrossRef]

30. Samsalee, N.; Sothornvit, R. Effects of drying methods on physicochemical and rheological properties of porcine plasma protein. *Kasetsart J. Nat. Sci.* **2014**, *48*, 629–636.

31. Ismoyo, F.; Wang, Y.; Ismail, A.A. Examination of the effect of heating on the secondary structure of avidin and avidin-biotin complex by resolution-enhanced two-dimensional infrared correlation spectroscopy. *Appl. Spectrosc.* **2000**, *54*, 939–947. [CrossRef]

32. Chantrapornchai, W.; McClements, D.J. Influence of NaCl on optical properties, large-strain rheology and water holding capacity of heat-induced whey protein isolate gels. *Food Hydrocoll.* **2002**, *16*, 467–476. [CrossRef]

33. Shao, J.J.; Zou, Y.F.; Xu, X.L.; Zhou, G.H.; Sun, J.X. Effects of NaCl on water characteristics of heat-induced gels made from chicken breast proteins treated by isoelectric solubilization/precipitation. *CyTA J. Food* **2016**, *14*, 145–153. [CrossRef]

34. Meng, X.; Peng, Z.; Jin, H.; Wu, D.; Feng, Y.; Cui, G. Study on gel properties of bovine plasma proteins. *Food Res. Dev.* **2012**, *33*, 157–160.

35. Croguennec, T.; Nau, F.; Brulé, G. Influence of pH and salts on egg white gelation. *J. Food Sci.* **2002**, *67*, 608–614. [CrossRef]

36. Zhang, Z.; Yang, Y.; Tang, X.; Chen, Y.; You, Y. Chemical forces and water holding capacity study of heat-induced myofibrillar protein gel as affected by high pressure. *Food Chem.* **2015**, *188*, 111–118. [CrossRef]

37. Boye, J.I.; Ismail, A.A.; Alli, I. Effects of physicochemical factors on the secondary structure of beta-lactoglobulin. *J. Dairy Res.* **1996**, *63*, 97–109. [CrossRef] [PubMed]

![foods logo] *foods*

MDPI

Perspective

Cheese Whey Processing: Integrated Biorefinery Concepts and Emerging Food Applications

Iliada K. Lappa [1,†], **Aikaterini Papadaki** [1,†], **Vasiliki Kachrimanidou** [1,2,*], **Antonia Terpou** [1], **Dionysios Koulougliotis** [3], **Effimia Eriotou** [1] and **Nikolaos Kopsahelis** [1,*]

[1]   Department of Food Science and Technology, Ionian University, Argostoli, 28100 Kefalonia, Greece
[2]   Department of Food and Nutritional Sciences, University of Reading, Berkshire RG6 6AP, UK
[3]   Department of Environment, Ionian University, Panagoula, 29100 Zakynthos, Greece
*   Correspondence: vkachrimanidou@gmai.com (V.K.); kopsahelis@upatras.gr (N.K.)
†   Equal contribution as first author.

Received: 22 July 2019; Accepted: 10 August 2019; Published: 15 August 2019

Abstract: Cheese whey constitutes one of the most polluting by-products of the food industry, due to its high organic load. Thus, in order to mitigate the environmental concerns, a large number of valorization approaches have been reported; mainly targeting the recovery of whey proteins and whey lactose from cheese whey for further exploitation as renewable resources. Most studies are predominantly focused on the separate implementation, either of whey protein or lactose, to configure processes that will formulate value-added products. Likewise, approaches for cheese whey valorization, so far, do not exploit the full potential of cheese whey, particularly with respect to food applications. Nonetheless, within the concept of integrated biorefinery design and the transition to circular economy, it is imperative to develop consolidated bioprocesses that will foster a holistic exploitation of cheese whey. Therefore, the aim of this article is to elaborate on the recent advances regarding the conversion of whey to high value-added products, focusing on food applications. Moreover, novel integrated biorefining concepts are proposed, to inaugurate the complete exploitation of cheese whey to formulate novel products with diversified end applications. Within the context of circular economy, it is envisaged that high value-added products will be reintroduced in the food supply chain, thereby enhancing sustainability and creating "zero waste" processes.

Keywords: food processing; integrated biorefineries; circular-economy; whey proteins; lactose esters; prebiotics; hydrogels; edible films; bacterial cellulose; carotenoids

1. Introduction

Design of integrated biorefinery concepts endeavors a consolidated approach to valorize all possible waste and by-product streams under the concept of circular economy. In line with this, it is a prerequisite to target the formulation of multitude products rather than a single product to secure sustainable processes. On top of that, the high-value products will need to find value-added end applications, whereas the food sector is deemed of high importance. Food products with added value will ultimately meet consumers' demands and confer possible health benefits.

Cheese whey constitutes a by-product of the dairy industry and refers to the liquid stream deriving from the transformation of milk into cheese, specifically from the process of agglomeration of casein micelles. Whey is mainly composed of water, but also contains around 50% of the milk solids [1,2]. The dry matter fraction retains most of the lactose (66–77%, *w/w*), 8–15% (*w/w*) of numerous types of globular proteins, along with 7–15% (*w/w*) of minerals salts [3]. The amount of whey generated relates to the amount of cheese production and also to the productivity based on the type of milk, whereby approximately 9 L of whey are obtained for every 1 kg of cheese produced [4,5].

Over the last decades, cheese whey is considered the most important pollutant of the dairy industry, associated with serious environmental hazards in the case that designated sustainable treatments are not applied. The major issue lies in the high organic load, mainly due to the high content of lactose but also to the occurrence of hardly-biodegradable proteins [6]. More specifically, chemical oxygen demand (COD) of cheese whey can vary from 50,000 to 80,000 mg/L, whereas biochemical oxygen demand (BOD) is around 40,000 to 60,000 mg/L [7]. The rapid consumption of oxygen in the soil caused from the breakdown of proteins and sugars present in whey poses a significant disposal problem, in line with the vast amounts of volumes generated. The global whey production in 2016 was estimated at 200 million t with an annual linear increase of 3% for the last 21 years [8].

Nonetheless, in the last decades, cheese whey characterization has been altered from waste to dairy side-stream product. Significant research has been conducted to mitigate viable and environmentally benign valorization alternatives for cheese whey, rather than just field disposal [9,10]. The high nutritional value of cheese whey [11,12] has induced the valorization of approximately 50% of residual whey [13–15] towards the generation of value-added products for food and chemical industries. Traditional uses of whey protein as a health promoter have been earlier reported, both in human and animal nutrition [16]. Furthermore, several process technologies and biotechnological approaches have also been developed to convert this by-product into a resource of valuable components or into an ample range of marketable beverages [17–19]. Scientific studies have demonstrated the nutritional and functional value of whey protein and have focused on developing a number of recovery methods via physicochemical processes [20–22]. Advanced technologies such as ultrafiltration and nanofiltration have enhanced the exploitation of whey streams [5,23,24]. Besides the implementation of these techniques, the deproteinized cheese whey constitutes a lactose-rich fraction, and still displays a BOD_5 >30 kg m^{-3} [25]. Equally, the obtained fraction exhibits a high organic pollutant that should be further treated or employed as onset material for valorization processes.

Microbial-based processes to convert cheese whey into valuable products have flourished as a potential route for biorefinery development. Fermentation processes could significantly decrease the organic load (lactose content), thus enabling an economical and feasible alternative utilization of cheese whey, thereby reducing the environmental impact. However, to the best of our knowledge, there are scarce studies performed to evaluate both whey fractions in the frame of an integrated and consolidated approach which could find applications within the food industry itself. Bioprocess integration is defined as the simultaneous incorporation of more than two-unit operations in a single process, thereby enabling the utilization of an organic-rich effluent like cheese whey to generate multiple additional products.

The purpose of this study is to elaborate an overview on the conversion of whey deriving from cheese manufacture to high value-added products. Recent developments and new insights in the processing and refining technologies for cheese whey exploitation are reported, while advanced approaches with special focus on food applications are taken into consideration. The overall aim of this article is to explore potential schemes that could be applied for this by-product, by introducing the concept of novel biorefineries. Novel and cost-effective exploitation concepts have emerged to be of paramount importance, thus potential research gaps are also identified by proposing holistic approaches of cheese whey valorization to formulate a multitude of end-products. Therefore, biorefining processes that implement the valorization of lactose and whey protein towards the formulation of high value-added products through enzymatic, microbial, and chemical methods are proposed. Ultimately, it is anticipated that novel and functional foods with enhanced properties will be the target end-products, allowing the reintroduction in the food manufacturing sector, within the concept of transitioning to a closed-loop circular economy.

2. Bioprocess Development Using Whey Lactose

The deployment of cascade bioprocesses, to foster a holistic approach for cheese whey valorization and mitigate its disposal, has gained significant scientific attention during the last years. Pharmaceutical and food industries exhibit potential market outlets for the lactose fraction deriving from cheese whey. Special attention is given in the production of added-value compounds as a result of enzymatic catalysis or microbial fermentations. In this context, the production of lactic acid, ethanol, microbial lipids, microbial biomass, single cell protein, poly-hydroxyalkanoates, enzymes, and endo-polysaccharides has been addressed in numerous studies dealing with the exploitation of whey lactose [13,26,27]. However, this article will focus on novel, promising, and not fully developed or extensively studied bioprocesses from whey lactose, targeting food applications in the context of functional food manufacture. In particular, the synthesis of lactose derivatives, mainly as novel and targeted prebiotic oligosaccharides and fatty acids esters, has lately attracted great interest [28–30]. Likewise, various food additives and functional components, such as biocolorants, medicinal mushrooms, spirulina, etc., can be produced through microbial fermentations. These aspects and other recent trends in the field of whey lactose upgrading are described in the following sections.

2.1. Enzymatic Bioprocesses

2.1.1. Galacto-Oligosaccharides

The prebiotic definition has been constantly evolving since the first definition in 1995 by Gibson and Roberfroid [31], being recently revised as "a substrate that is selectively utilized by host microorganisms conferring a health benefit" [32]. Prebiotic oligosaccharides are non-digestible compounds, varying in the composition and configuration of monosaccharide residues and the type of glycosidic linkages. Prebiotics confer beneficial effects on human health, primarily by modifying the indigenous colonic gut microbiota [33]. Prebiotic oligosaccharides can be found in fruit, vegetable, dairy, and seafood processing by-products, while they can also be enzymatically synthesized [34]. Galacto-oligosaccharides (GOS) and lactulose are well established and recognized prebiotics, based on their health-promoting effects, including immunomodulation, lipid metabolism, mineral absorption, weight management, and obesity-related issues, among others [35].

Galacto-oligosaccharides are non-digestible, galactose-containing oligosaccharides with a unit of terminal glucose in the form Glu α1–4(β Gal 1–6)n, showing a degree of polymerization (DP) ranging from 2 to 8–9 [36,37]. GOS are produced by lactose, through the transgalactosylation action of β-galactosidase (EC 3.2.1.23), yielding a mixture of oligosaccharides, mono- and disaccharides, with a high range of linkages, mainly β1–4 and β1–6, but also β1–3 and β1–2 [38]. GOS occur naturally in the milk of animals and humans at low concentrations, but they are mainly produced by chemical glycosylation or enzymatic routes to meet market demands. The worldwide market size of GOS was estimated at $703.8 million in 2017, and it is anticipated to increase significantly by 2025, following the constantly-rising demand for the consumption of dietary supplements [39]. The prebiotic effect of GOS has been widely demonstrated during in vitro animal and human studies (including studies with infants). GOS have been recognized as safe (GRAS) in the United States and are characterized as foods for specific health use (FOSHU) in Japan, where they have been applied in a spectrum of end-products such as sweeteners, bulking agents, and sugar substitutes [38,40,41]. Thereby, GOS are mostly applied to infant formula products, aiming to formulate products similar to human milk composition, but also in beverages, meal replacers, flavored milk, and confectionery products (e.g., bread). Their incorporation into food products has been regulated in many countries, which are using GOS as functional food ingredients, whereas in Europe, they are under pre-screening evaluation from the European Food Safety Authority (EFSA) [41,42]. Apart from the food industry, GOS have found applications in the animal feed, cosmetic, and pharmaceutical industries.

Another lactose-derived prebiotic is lactulose (4-O-β-D-galactopyranosyl-D-fructose), a non-digestible synthetic disaccharide, comprising of fructose and galactose [30]. Lactulose has been marketed mainly as a medical product [37], finding many applications in food products such as milk for bottle-fed babies to adjust the composition of their colonic microbiota [43].

The chemical route is a common method for the production of GOS and lactulose. The main drawbacks of the chemical synthesis are the requirement of catalysts and chemicals, the low specificity, and the production of undesirable compounds. However, many companies have employed enzymatic synthesis, as it offers several advantages, including the requirement of non-purified substrates, selectivity, mild reaction conditions, and lower downstream operation costs [30]. The enzymatic production and the final configuration of GOS, in terms of molecular weight distribution and linkages, are affected by various factors, including the concentration of lactose and water during the reaction and the source of the enzyme employed. The biocatalyst β-galactosidase can be obtained from several microbial sources, including *Kluyveromyces lactis, Bacillus circulans, Bifidobacterium bifidum, Aspergillus oryzae,* and *Streptococcus thermophiles* [44]. During GOS synthesis with β-galactosidases, lactose acts both as donor and acceptor of the transgalactosylated galactose [45], whereas during lactulose synthesis, lactose is the galactosyl donor and fructose acts as the acceptor. However, a mixture of lactulose and GOS is produced during lactulose synthesis, as lactose and fructose are simultaneously present in the reaction medium and, thus, can act as acceptors and their production ratio depends on the process conditions [46]. Likewise, GOS purification steps are essential when food applications are targeted. Hernández et al. [36] evaluated several fractionation techniques, showing the potential of yeast treatment to obtain high purity GOS, compared to diafiltration and activated charcoal [36].

The use of whey lactose has been suggested as an alternative substrate for the enzymatic synthesis of potential prebiotics, leading to a more sustainable and competitive process within the concept of bioeconomy. Although there are many studies reporting GOS production using primarily pure lactose [28], research has been also conducted using whey lactose as substrate. Lactose from whey can be obtained via crystallization of a supersaturated solution [41]. Wichienchot and Ishak [47] suggested that lactose derived from cheese whey is a potential source for GOS and lactulose production [47]. Splechtna et al. [48] found that GOS production, catalyzed by a β-galactosidase of *Lactobacillus* sp., was reduced compared to buffered lactose substrate, whereby GOS yield was 28% of total sugars. However, higher yields have been reported by other studies [48]. Das et al. [49] reported 77% GOS production from whey lactose by employing β-galactosidase from *Bacillus circulans* [49]. High GOS production (53.45 g/L) has been also produced from lactose-supplemented whey, catalyzed by the β-galactosidase of *Streptococcus thermophilus* [35]. Díez-Municio et al. [50] indicated that cheese whey is a suitable material for the synthesis of the trisaccharide 2-α-D-glucopyranosyl-lactose [50]. The authors mentioned a yield of 50% of the initial amount of lactose, under the optimum reaction conditions. A co-reaction was performed using bovine cheese whey and tofu whey as lactose and sucrose sources, respectively, for the production of 80.1 g/L lactosucrose. This approach allowed the simultaneous utilization of two by-products which resulted in a very high productivity of 40.1 g lactosucrose /L/h [51]. In another study, a continuous reaction was performed using β-glucosidase from *Kluyveromyces lactis,* achieving a maximum yield of 31% oligosaccharides in a pilot plant scale UF-hollow fiber membrane reactor [52]. Lower yields up to 11.3 % of GOS have been obtained in other studies using different types of cheese whey (sweet whey, acid whey) [53]. Overexpression of β-galactosidase from *S. thermophilus* in a food grade *L. plantarum* strain resulted in the production of 50% of GOS using 205 g/L of lactose derived from whey [54], thus indicating an efficient valorization route for whey lactose.

For the commercial production of GOS, Nestle company has developed a procedure using partially demineralized sweet whey permeate. Initially, whey is concentrated and then β-glucosidase produced from *A. oryzae* is added and the reaction is stopped through heat inactivation of the enzyme [42].

Scott et al. [55] performed a techno-economic analysis to evaluate the production of whey powder and lactose as market outlet for subsequent GOS production [55]. The plant capacity, along with the current prices of whey powder and lactose, were closely affected with the profitability of the complete process. Nonetheless, the authors suggested that the bioprocess and restructuring of the plant could become more robust if the price of whey powder rises [55]. On the other hand, the development of integrated cheese whey biorefineries towards the production of added-value products, from both lactose and whey protein streams, could exploit the full potential of cheese whey, including process optimization and downstream recovery that could be annexed to other bioprocesses. Added-value products from whey protein, and the possibility to configure cascade bioprocessing for cheese whey, will be elaborated in the following sections, proposing the development of robust integrated scenarios.

2.1.2. Lactose Fatty Acid Esters

Sugar esters are odorless, non-toxic, and biodegradable compounds of high importance for the food industry [30]. The most common sugar esters derive from sucrose, with an estimated global market of $74.6 million in 2020 [56]. Although, lactose esters have not been extensively studied, they have found several applications within the food, cosmetic, and pharmaceutical industries [57]. These sugar esters demonstrate excellent emulsifying and stability properties in food products, whereas they may be applied as low-fat alternatives. Additionally, they present antimicrobial activity against many foodborne pathogens, as well as medicinal properties such as anticancer activity [30,57].

Chemical synthesis of lactose esters is the most common route for their production. The main drawback of the chemical lactose esterification is the production of non-stereospecific esters [57]. The use of enzymes, such as lipases, esterases, and proteases, affects the reaction selectivity due to their regiospecifity [57]. Among all enzymes, lipases have attracted significant interest due to their stability during several batch reactions at high temperatures and their ability to utilize different substrates [57–61].

Lactose ester production has been studied since 1974 [62]. An extensive review for lactose ester production through enzymatic catalysis demonstrated that enzymes from various microbial sources—e.g., *Candida antarctica*, *Mucor miehei*, and *Pseudomonas cepasia*, among others—can be utilized, entailing high yields. More specifically, lipases have achieved yields up to 89%, whereas the protease from *Bacillus subtilis* reached the highest yield (96%), at mild temperature conditions (45 °C) [63,64]. During sugar ester synthesis, fatty acid vinyl esters are utilized as acyl donors. Since vinyl esters are expensive and result in unstable by-products (vinyl alcohols), Enayati et al. [65] replaced them with fatty acids, such as lauric acid and palmitic acid, which yielded high lactose ester synthesis (93%) [65]. This method could be further developed by employing renewable resources with a high content of free fatty acids, such as fatty acid distillates. For instance, palm fatty acid distillate has been successfully valorized towards polyol ester production using a commercial lipase [61]. Even though lactose esters have been recognized for their superior properties and as attractive substitutes of synthetic surfactants [65,66], only pure lactose has been employed for their production until now [57].

2.2. Microbial Bioprocesses

2.2.1. Food Biocolorants and Aroma Compounds

Carotenoids are considered one of the most important groups of natural pigments, exhibiting numerous biological functions. Carotenoids are characterized by their antioxidant activity and their exceptional health benefits on human health, such as the reduction of cardiovascular diseases, anti-diabetic, anti-cancer, and anti-inflammation activities [67–69]. Humans are not able to synthesize carotenoids; thus, their uptake can be only performed via the consumption of carotenoid-rich food products. The most commercially important carotenoid is β-carotene, followed by lutein and astaxanthin. Likewise, β-carotene is widely applied as a food supplement, acting as provitamin A, and as a coloring agent in food products, such as butter, margarine, cheese, confectionery, ice cream, juices, other beverages, etc. [70,71]. Natural astaxanthin has gained industrial interest as it presents significantly higher antioxidant activity than the respective counterpart made via the chemical route [72]. Astaxanthin is widely utilized in salmon aquaculture and as a dietary-supplement for human consumption [71,73].

Carotenoids were initially extracted from plants, but they are currently produced primarily through chemical synthesis. Natural origin carotenoids can be obtained only through plant extraction or biotechnologically. The fermentative production of carotenoids has been well-investigated using various carbon sources, such as glucose, sucrose, and xylose, among others; however, the interest has been shifted to the use of low-cost substrates, aiming to reduce the high production cost. In this context, there is a growing interest for the development of bioprocesses using renewable resources as alternative carbon sources [71]. The fungus *Blakeslea trispora*, as well as many yeast species belonging to the genera of *Rhodosporidium* sp. *Rhodotorula* sp. and *Phaffia rhodozyma*, have been studied for carotenoid production using low-cost substrates [71,74]. Usually, most of them produce a mixture of carotenoids consisting of β-carotene, torulene, torularodine, and γ-carotene [71]. In the case of *P. rhodozyma*, the carotenoid mixture primarily comprises astaxanthin [75]. The microalgae *Haematococcus pluvialis* also constitutes a rich source of astaxanthin, thus presents the highest potential for astaxanthin production [73].

Among renewable resources, cheese whey has emerged as a promising candidate for carotenoid production, however only a few studies are found in the literature. Cheese whey, or deproteinized cheese whey, has been utilized for the production of carotenoids using various microorganisms. Nevertheless, most of them are lactose-negative species, thus in many cases, enzymatic hydrolysis of deproteinized cheese whey is carried out prior to fermentation. Table 1 summarizes all the results with respect to carotenoid production from cheese whey up-to-date. Evidently, *B. trispora* demonstrates the highest yields among all microorganisms. The highest carotene production of 1620 mg/L with an intracellular yield of 222 mg/g was reported by Roukas et al. [76]. The fermentation was carried out in a bubble column reactor using deproteinized, hydrolyzed cheese whey. This is among the highest values achieved with agro-industrial by-products, indicating that cheese whey might be one of the most promising renewable resources for the commercial production of carotenes. Table 1 also shows that the selection of the proper microbial strains can lead to the production of specific carotenoid types. For instance, carotenoids rich in canthaxanthin can be obtained by the bacterium *Dietzia natronolimnaea* [77]. Apparently, astaxanthin production has not been studied yet implementing cheese whey as substrate. This can be attributed to the fact that *P. rhodozyma* cannot assimilate lactose and galactose [78], whereas among several sugars, lactose results in the lowest astaxanthin production using the microalgae *Chlorella zofingiensis* [79].

Carotenoid production—and particularly the proportions of individual carotenoids—correlate to several factors (e.g., the addition of surfactants and vegetable oils) along with culture conditions, (e.g., aeration rate) [76,80–82]. Interestingly, these studies have suggested the potential combination of oil by-products with cheese whey to foster a promising and circular valorization of food by-products for carotenoid generation.

Table 1. Carotenoid production from various microorganisms through fermentation in cheese whey (CW).

Microorganism	Supplementation of CW Medium	Composition of Total Carotenoids	Concentration (mg/L)	Yield (mg/g) [1]	Reference
Blakeslea trispora ATCC 14271 & ATCC 14272	Tween 80, Span 80, β-ionone	β-carotene, γ-carotene, lycopene	1620.0	222.0	[76]
Blakeslea trispora ATCC 14271 & ATCC 14272	Tween 80, Span 80, β-ionone	β-carotene, γ-carotene, lycopene	1360.0	175.0	[83]
Blakeslea trispora ATCC 14271 & ATCC 14272	Tween 80, Span 80, vegetable oils	β-carotene, γ-carotene, lycopene	~672.0	16.0	[82]
Blakeslea trispora ATCC 14271 & ATCC 14272	Tween 80, Span 80, vegetable oils, antioxidants and other nutrients	β-carotene	350.0	11.6	[84]

Table 1. *Cont.*

Microorganism	Supplementation of CW Medium	Composition of Total Carotenoids	Concentration (mg/L)	Yield (mg/g) [1]	Reference
Blakeslea trispora ATCC 14271 &ATCC 14272	Tween 80, Span 80, vegetable oils	N.S. [2]	376.0	8.0	[85]
Mucor azygosporus MTCC 414	Soluble starch	β-carotene	3.5	0.38	[86]
Rhodotorula mucilaginosa NRRL 2502		N.S. [2]	70.0	29.2	[80]
Rhodotorula mucilaginosa CCY 20-7-31		β-carotene	11.3	0.38	[87]
Rhodotorula glutinis CCY 20-2-26		β-carotene	51.2	1.48	[87]
Rhodotorula rubra GED5 co-culture with *Kluyveromyces lactis* MP11		Torularhodin, β-carotene, torulene	10.2	0.42	[88]
Rhodotorula glutinis 22P co-culture with *Lactobacillus helveticus*		β-carotene, torularhodin, torulene	8.09	0.27	[89]
Rhodotorula rubra GED8 co-culture with *Lactobacillus bulbaricus, Streptococcus thermophilus*		β-carotene, torulene, torularhodin	13.1	0.50	[90]
Sporidiobolus salmonicor CBS 2636		N.S. [2]	0.91	0.25	[91]
Sporobolomyces roseus CCY 19-4-8		β-carotene	29.4	2.89	[87]
Dietzia natronolimnaea HS-1		canthaxanthin (2.87 mg/L)	3.06	~0.9	[77]

[1] Yield expressed as milligrams of carotenoids per gram of dried biomass; [2] N.S.: Not specified.

Likewise, pulcherrimin is a red pigment belonging to cyclodipeptides, characterized for its strong biological properties, including antibacterial, antifungal, antitumoral, and anti-inflammatory activities [92]. Türkel et al. [93] mentioned that microorganisms producing pulcherrimin can be effectively used as biocontrol agents against various postharvest pathogens causing fruit and vegetable spoilage, due to the antimicrobial activity of the pigment [93]. Pulcherrimin production has been identified as a metabolite of the yeast *Metschnikowia pulcherrima*, but it has been poorly investigated until now [94,95]. *M. pulcherrima* is able to metabolize various carbon sources, including galactose and glucose, but it cannot hydrolyze lactose [95]. This indicates that hydrolyzed whey lactose could be employed as fermentation feedstock for pulcherrimin production. Alternatively, whey lactose could also be utilized by *Bacillus licheniformis*, which is able to assimilate lactose and has presented the highest pulcherrimin production of 331.7 mg/L under optimized culture conditions [92].

Similarly, flavor and aroma compounds constitute another essential category for the food industry. Those compounds are widely used in order to manufacture attractive products to consumers.

Chemical synthesis is an inexpensive method for the production of aroma compounds; however, the derived products cannot be applied in foods. On the other hand, the traditional extraction of aroma compounds from plants exhibits disadvantages regarding low yields and high production cost. In this context, fermentation processes could provide an alternative way for the production of natural aroma compounds. Few studies have focused on cheese whey valorization for the production of fragrances. Several yeast strains were isolated and screened for the production of 2-phenylethanol, an aroma compound found in rose petals, using a whey medium supplemented with sugar beet by-products (molasses, thick juice, or sludge) and L-phenylalanine as a precursor. Among all strains, the highest concentration of 3.3 g/L was achieved by a *Saccharomyces. cerevisiae* yeast strain [96]. The strain *Metschnikowia pulcherrima* is also a promising producer of 2-phenylethanol. Currently, there is not any published study using whey; however, utilization of simulated grape juice medium resulted in significant production of 2-phenylethanol (14 g/L) [97]. Other aroma compounds, including 2-phenylethanol, have been identified at low concentrations in a whey-glucose substrate fermented by the yeast *Wickerhamomyces pijperi*. In total, twelve aroma compounds such as isobutanol, isoamyl alcohol, 2-phenylethanol, acetaldehyde, ethyl acetate, propyl acetate, isobutyl acetate, isoamyl acetate, ethyl butyrate, ethyl propionate, ethyl hexanoate, and ethyl benzoate have been determined [98].

2.2.2. Bacterial Cellulose

Bacterial cellulose is a microbial polysaccharide presenting improved water holding capacity, hydrophilicity, high degree of polymerization, mechanical strength, crystallinity, porosity, and purest fiber network, compared to plant cellulose. Several food applications have been developed for bacterial cellulose, since it has been characterized as a "generally recognized as safe" (GRAS) food by the US Food and Drug Administration (FDA). It has already been applied in ice-creams as a rheology modifier, in confectionery products as a fat replacer, as artificial meat for vegetarian consumers, as a stabilizer of emulsions, or as an immobilization carrier of probiotics and enzymes [99].

Bacterial cellulose is synthesized from several *Acetobacter* species. *Gluconacetobacter xylinus* (formerly known as *Acetobacter xylinum*) is one of the most studied species because of its ability to produce high bacterial cellulose concentrations using various substrates [100]. The implementation of inexpensive renewable resources and agro-industrial wastes as fermentation media could alleviate the high cost for bacterial cellulose production that hinders large scale manufacture. Bacterial cellulose production has been previously studied using by-products deriving from biodiesel and food industries, such as sunflower meal, glycerol, confectionery wastes, citrus by-products, grape pomace, and discarded currants, among others [100–104]. High yields (up to 15.2 g/L) of bacterial cellulose were obtained from synthetic sucrose, glucose, and fructose media [105–107]. Agro-industrial substrates such as molasses, fruit juices, or aqueous extracts from citrus residues resulted in production yields of up to 7.8 g/L [101,105]. In the case of lactose utilization, there are only a few studies dealing with bacterial cellulose production (Table 2). Tsouko et al. [100] demonstrated that synthetic lactose was not efficiently metabolized from *Komagataeibacter sucrofermentans* DSM 15973, yielding up to 1.6 g/L bacterial cellulose [100]. Likewise, Mikkelsen et al. [108] reported a final bacterial cellulose production of 0.1 g/L by *G. xylinus* ATCC 53524 grown on galactose [108]. Similarly, other studies have agreed that cheese whey does not support significant bacterial cellulose production by *A. xylinum* 10821, *A. xylinum* 23770 [109], and isolated from Kombucha tea *G. sacchari* [110]. This could be attributed to the fact that the gene that encodes β-galactosidase is not expressed by bacterial cellulose producers. Battad-Bernardo et al. [111] produced a mutant of *A. xylinum* by inserting *lacZ* gene, thus allowing the hydrolysis of lactose [111]. The mutant strain was able to produce 1.82 g/L bacterial cellulose in a whey-based substrate [111]. Another approach lies in the pre-treatment of cheese whey through enzymatic catalysis. In this way, Salari et al. [112] improved bacterial cellulose production (3.55 g/L) by *G. xylinus* PTCC 1734 using an enzymatically-hydrolyzed cheese whey [112]. In another study, around 5.4 g/L of bacterial cellulose was produced by the isolated *Gluconacetobacter. sucrofermentans* B-11267 strain on untreated cheese whey [113].

Table 2. Bacterial cellulose (BC) production using lactose or lactose derivatives.

Microorganism	Carbon Source	BC (g/L)	Reference
K. sucrofermentans DSM 15973	Synthetic lactose	1.6	[100]
G. xylinus ATCC 53524	Synthetic galactose	0.1	[108]
A. xylinum 10821	Cheese whey	0.04	[109]
A. xylinum 23770	Cheese whey	1.13	[109]
G. sacchari	Cheese whey	0.15	[110]
A. xylinum mutant	Cheese whey	1.82	[111]
G. xylinus PTCC 1734	Hydrolyzed cheese whey	3.55	[112]
G. sucrofermentans B-11267	Cheese whey	5.4	[113]

2.2.3. Functional Food Additives

Development of functional food ingredients is of paramount significance for the food industry, driven by the rising demand to manufacture food products that confer health benefits. Among others, several research studies have been conducted focusing on the production of value-added compounds from mushrooms and microalgae. Still, the full potential of cheese whey was not fully exploited in these cases.

Mushrooms are known for their exceptional functional properties, primarily due to their polysaccharide content [114]. Mushroom fruiting bodies contain a significant amount (35–70%) of non-digestible—and to lesser extent, digestible—carbohydrates. Chitin, β-glucans, glucose, mannitol, and glycogen are the main component carbohydrates [115]. These components serve as the dietary fiber fraction, found mostly in the fungal cell wall, and possess many beneficial effects on human health, including antitumor, hepatoprotective, antimicrobial, prebiotic, antioxidant, hypoglycemic, and hypolipidemic activity [114]. Various mushroom species have been studied for the production of polysaccharides [116,117]. However, the mushrooms belonging to the genus of *Pleurotus* sp., *Ganoderma lucidum,* and *Lentinula edodes* (shiitake mushroom) have been most extensively studied.

Many *Pleurotus* species have demonstrated biological effects, such as immunostimulating and antitumor activity. More specifically, *P. ostreatus* and *P. eryngii* contain several water-soluble and non-soluble β-glucans with prebiotic properties [118]. Recently, Velez et al. [119] showed that the mycelium of *P. djamor* was rich in ergosterol and β-glucans, presenting also high antioxidant activity when grown in cheese whey supplemented with sodium selenite [119]. It was suggested that the lactose-free mycelium, rich in bioactive compounds, is an appropriate supplement for consumers with lactose intolerance. In the case of cheese whey fermentation by *P. sajor-caju*, mycelium biomass was found rich in carbohydrates (arabinose, mannose, and N-acetylglucosamine) and proteins (39.2%) containing high amounts of essential amino acids, such as lysine, leucine, threonine, and phenylalanine [120]. Likewise, *P. osteatus* exhibited higher contents of water-soluble polysaccharides and trace elements, such as calcium, phosphorus, potassium, sodium, and magnesium, when cultivated in whey permeate rather than in synthetic medium [121]. A previous study has shown that cheese whey could also be efficiently utilized in solid state fermentations by *P. ostreatus* [122].

Lentinula edodes produces a polysaccharide, namely lentinan, which is the most studied immunomodulating polysaccharide and commercially available in pharmaceutical products. Glycoproteins from this mushroom have been characterized for their antitumor activity [123]. Few reports of *Lentinula edodes* cultivation in cheese whey. have shown that the mycelium is rich in water-soluble polysaccharides and minerals (calcium and potassium) showing also high antioxidant capacity, which suggests that dairy by-products could be utilized as a growth substrate for the cultivation of *L. edodes* [124–126].

Few studies have included the optimization of culture conditions using cheese whey for the production of polysaccharides from *Ganoderma lucidum,* indicating its ability to be used as an alternative fermentation medium [127–129]. This mushroom has been characterized as "the mushroom of immortality" in Asian countries, because of the existence of 400 different bioactive compounds in mycelia and fruiting body. It is recognized as an alternative adjuvant in the treatment of leukemia, carcinoma, hepatitis, and diabetes, and its bioactive compounds consist of triterpenoids, polysaccharides, nucleotides, sterols, steroids, fatty acids, and proteins, among others [130]. Recently, *G. lucidum* polysaccharides were applied to the production of microcapsules using whey proteins as wall material. This product showed better stability and controlled release ability [131]. This is a representative paradigm of polysaccharide production from medicinal mushrooms, and subsequent product development through encapsulation, in the framework of an integrated cheese whey biorefinery.

Another less studied, yet highly valued, edible mushroom is *Morchella* sp. It is a delicious and expensive mushroom containing heteroglycans with antitumor and hypoglycemic properties [118]. Many species of *Morchella* sp. have demonstrated high polysaccharide content after cultivation on agro-industrial substrates [132]. Nevertheless, the use of cheese whey as substrate for *Morchella* production is still unexplored. Kosaric and Miyata reported, for the first time in 1981, that several *Morchella* species, such as *M. crassipes*, *M. esculenta*, *M. deliciosa*, *M. rotunda*, and *M. angusticeps*, were able to grow on partially deproteinized cheese whey. *M. crassipes* produced the highest biomass of 20 g/L, which contained 45% protein, many essential amino acids, and high proportion of unsaturated fatty acids, such as linoleic acid (55.7%) and oleic acid (13%) [133].

In a similar way, microalgae species are known for their ability to accumulate high quantities of polysaccharides, proteins, polyunsaturated fatty acids, and carotenoids—thus, they are considered superior candidates for food supplements [134]. *Spirulina* and *Chlorella* are blue-green microalgae species exhibiting various health benefits, including antiviral, anti-inflammatory, and antitumor properties [135]. Cultivation of microalgae in cheese whey has been considered a feasible alternative for cost-effective production of microalgal biomass production [136]. More specifically, the mixotrophic culture of *Chlorella vulgaris* in a medium supplemented with hydrolyzed cheese whey demonstrated improved biomass production. This was attributed to the presence of growth-promoting nutrients in cheese whey [136]. *Spirulina platensis* presented higher carbohydrate, carotenoid, and chlorophyll contents when the substrate was supplemented with cheese whey [137,138]. Girard et al. [139] substituted a significant quantity (40%, *v/v*) of the basal medium with whey permeate, which resulted in higher biomass production by the fresh water green algae *Scenedesmus obliquus* under mixotrophic conditions [139]. The significance of *Scenedesmus obliquus* is on its enzymatic extracts, which contain several amino acids essential for human diet, presenting also antioxidant and antiviral activities [140,141].

3. Whey Proteins: Research Insights and Trends

The increasing global demand for natural ingredients in food manufacturing has led to significant research interest on whey proteins (WP) to manufacture products with desirable characteristics. Whey proteins exhibit physicochemical properties resulting in enhanced texture and quality of end-products, regarding structural and rheological functions [142,143]. Surface-active components, texture modifiers, foaming and gelling agents, thickening agents, emulsifiers, and other bioactivities, among others, indicate targeted application of WP as active ingredients [144–146] (Figure 1).

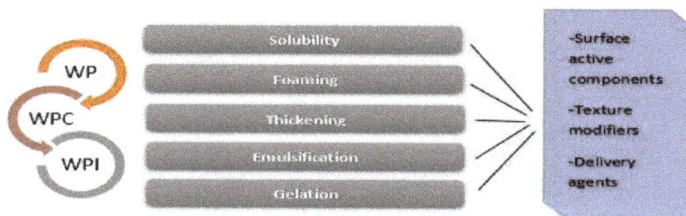

Figure 1. Technological functions of whey protein (WP), whey protein concentrate (WPC), and whey protein isolate (WPI) in food applications.

WPs can develop macro-, micro-, and nano-structures with numerous promising food applications, such as vehicle carrying for various bio-compounds, flavors, or nutrients (Figure 2). Hence, the emerging development of WP processing techniques could enable further applications that will be directed towards value-added products. In an analogous approach, the following sections will elaborate the most promising applications of cheese whey proteins. The development of edible films and coatings, nanoparticles like hydrogels, and the production of whey protein bioactive peptides as potential nutraceuticals will be further described. Within this context, novel approaches on WP implementation could foster the spectrum of end-use applications that could be incorporated in the development of a holistic process for cheese whey valorization.

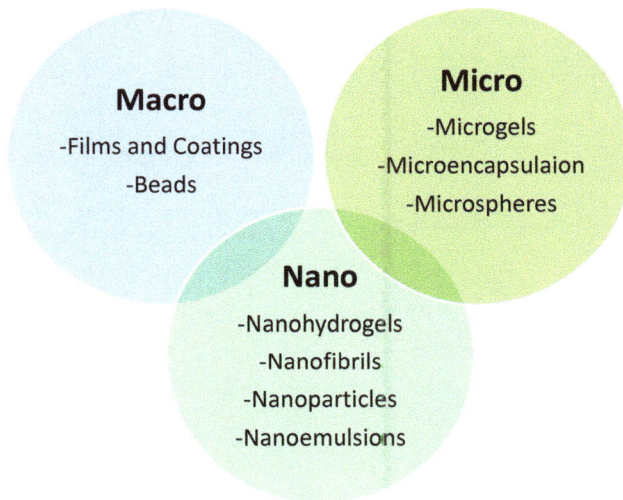

Figure 2. Whey protein systems used as delivery vehicles for bioactive ingredients in food.

3.1. Edible Films and Coatings

3.1.1. Recent Strategies for Improved Technical and Functional Properties

The flourishing demand for eco-friendly active packaging has stimulated research on bio-based packaging. The use of WP for the formation of edible films and coatings has drawn scientific interest, as they are produced from an abundant and renewable material compared to the synthetic counterparts. Among biopolymers used to fabricate edible films, WP exhibits diverse and distinctive technological properties. More specifically, WP can form transparent films and coatings with improved mechanical and barrier properties compared to polysaccharide-based films, indicating them as potential candidate for numerous applications (e.g., high barrier properties like oxygen and volatiles under low moisture conditions) [147]. WPs are usually employed in food application as whey protein concentrates (WPCs) and whey protein isolates (WPIs), induced by the technological advances of whey processing, that convey enhanced functionality (Table 3).

Table 3. Edible films formation from whey protein isolate (WPI) and whey protein concentrate (WPC) and their functional features.

Substrate	Promoting Compound	Functionality	Reference
WPI	Almonds, walnut oil	Water barrier improvement	[148]
	β-cyclodextrin/eugenol, carvacrol	Antimicrobial component delivery	[149]
	Lysozyme	Antimicrobial component delivery	[150]
	Montmorillonite nanoplatelets	Oxygen barriers improvement	[151]
	Montmorillonite clay nanoparticles	Thermal stability, water vapor permeability	[152]
	Nanocrystalline cellulose, transglutaminase	Improved mechanical properties	[153]
	Oat husk nanocellulose	Enhanced tensile strength, solubility, decreased elongation at break and moisture content, decreased transparency and water vapor permeability	[154]
	Pullulan, montmorillonite	Improve the mechanical properties, thermal properties, and water resistance	[155]
	Sodium laurate-modified TiO_2 nanoparticles	Water vapor permeability decreased, tensile strength increase, decreased transparency	[156]
	Starch	Water vapor permeability, microstructure	[157]
	Zein	Enhanced water solubility and heat-sealablity	[158]
	Zein nanoparticles	Improved moisture barrier and mechanical properties	[159]
WPC	Cinnamon essential oil	Antimicrobial	[160]
	Glucerol, pullulan, beeswax	Improved color indices, diminished water solubility and water vapor permeability, and increased tensile strength	[161]
	Immunoglobulins	Increase stickiness, adhesion, and tensile strength of the films	[162]
	Liquid smoke	Antimicrobial/improved mechanical properties	[163]
	Montmorilonite, lycopene	Antioxidant activity and UV-vis light protection/mechanical properties improvement	[164]
	Rosmarinic acid, carnosol, carnosic acid		
	Sodium alginate, pectin, carrageenan, locust been gum/*L. rhamnosus*	Enhanced survival during drying and storage, reduced film water vapor permeability	[165]
	Sunflower, beeswax	Water vapor permeability	[166]

Incorporation of different additive compounds on whey protein-based films to improve their natural, technical and functional properties has lately attracted significant attention [167]. For instance, immunoglobulins (Ig) incorporated into whey protein films improved adhesion and strength [162], but also yielded more transparent and clear films. Moreover, the authors stated that embedding Ig in whey protein matrices resulted in the protection from rapid proteolysis, thereby sustaining their activity in the gastrointestinal track (GIT). Other modifications have been also used to improve whey protein films, including, for instance, the inclusion of lipid components to improve the moisture ability of such films. Almond and walnut oils were employed, along with WPI, leading to a reduction of the surface hydrophilic character of films [148]. Sunflower oil was also used with WPC, resulting in a reduction of water vapor permeability [166]. An emerging strategy to further enhance technical and functional properties of these films, thus improving the compatibility of polymers, lies in the incorporation of nanomaterials [152,156].

Likewise, the use of edible-coated nanosystems has also been considered as a novel approach for food preservation. Findings have lately suggested that novel WPI-based nanocomposites can be part of multilayer flexible packaging films, thus holding great potential to even replace well-established fossil-based packaging materials to support certain mechanical properties during storage [151,153,155,163]. Alternative biomaterials, have also been recently proposed for potential uses in foodstuff applications, via the production of nanocomposites from WPCs activated with lycopene and montmorillonite nanoparticles [164,168]. In another work, WPI nanocomposite films properties were reinforced with oat husk nanocellulose [154]. Evidently, novel perspectives are encountered in the development of novel packaging materials.

Notwithstanding, along with the increase of whey protein edible film generation, it is crucial to overcome specific disadvantages of these films in terms of mechanical features and moisture barrier properties. This could be alleviated by blending films and coatings with various plasticizers. As an example, Basiak et al. (2017) used various starch/whey proteins mixtures and studied their effect on transport properties of the produced films [157].

Likewise, several studies have suggested the use of plasticizers and crosslinking on the formulation of films, as a potential technology for novel food packaging, to improve water resistance, mechanical and barrier characteristics [152], while diversified blends, using different ratios of pullulan or sugars like trehalose, have been also recently proposed as plasticizers of whey protein films [161,169].

3.1.2. Delivery Agents of Bioactive Compounds

The deployment of specific active compounds with antimicrobial or natural antioxidant features, into the matrix of whey protein isolate formation, remains a field of significant importance [170,171]. Improved quality and safety control, along with extended shelf life, of the products, are among the main advantages reported by the use of such complex films, implying their potential application for food wrapping [150,160,172].

Another evolving aspect for bioactive whey protein-based films and coatings was presented via the incorporation of functional bacteria. The use of edible films and coatings as carriers of living microorganisms constitutes a challenge. Microorganisms should remain in high concentrations to exert beneficial effects (antimicrobial or probiotic) without affecting the mechanical or sensory properties of the product. Novel approaches using whey edible films and coatings have resulted in enhanced cell survival. WPC was evaluated, along with several selected biopolymers, as a potential vehicle to investigate *L. rhamnosus* GG survivability [165]. To date, only a few reports exist about probiotic activity in edible films and coatings from whey [164]. The possibility of implementing whey protein formations as a carrier matrix for viable probiotics could potentially result in better survival rates during storage and consumption, thereby promoting novel food applications.

3.2. Whey Protein Hydrogels

3.2.1. Formulations and Structural Characteristics

Whey proteins have the ability to form polymeric three-dimensional networks, including hydrogels systems [173,174], and further combine them into nanoparticles. For instance, β-LG, which is the main component of whey protein isolate, is also the responsible particle for the main functional properties. These hydrogels are considered as unique delivery systems, since β-LG nanoparticles are able to bind to hydrophobic compounds. The gel formation process of WP starts when proteins are partially denaturated above critical temperature, leading to the formation of the three-dimensional molecule structure.

Whey protein gels can be temperature-induced by either cold- or heat-set mechanisms [175,176], acid-induced [177], or even enzyme-induced [143,156]. Heat-induced gelation of whey protein is irreversible; therefore, gels are being extensively studied as potential candidates for the preparation of fluid gels without additives [178,179]. Recent studies have reported the preparation of whey protein aggregates through protein crosslinking and building blocks of cold-set gels, thus protecting substances from high temperatures [180,181]. Cold gelation is emerging as a rising method to formulate whey protein microgels [174]. This characteristic renders them as promising candidates to design functional products, where heat- or acid-sensitive nutraceuticals are encapsulated.

Structural characterization of hydrogels is crucial, in order to evaluate their potential in food applications. The rheology of whey protein isolate and casein micelles (MC) mixtures upon heating has been evaluated, indicating that WPI binds to MC and strengthens the junctions of the MC network [176]. Likewise, different mixtures, including starch, rice, or other polysaccharides, or other proteins combined with WPI are currently being studied, aiming to improve the structural and functional properties (e.g., strength, viscosity) of protein solutions [176,182].

The stability and strength of the protein–gel network is evidently associated with the designated applications in food and biomaterials. Characterization of WP gelation profiles has been recently studied; however, the main challenge remains in the production of effective WP microgel systems to overcome brittleness and susceptibility to syneresis [183]. Research is focusing on the ability of WP to form hydrogels that entail specific structural and sensory characteristics for targeted food products like yoghurt, ice cream, bakery products, desserts, and meat products [182]. Understanding the interactions of whey with other biopolymers is crucial in the sense of novel functional food properties. Therefore, the interactions of mixed systems of whey with other biopolymers—such as pectin, κ-carrageenan, xanthan, and basil seed gum—have been studied [184–187]. These synergistic interactions, leading to the production of stronger gels, could be beneficial in many food formulations such as dairy and dessert products.

3.2.2. Emerging Techniques for Food Applications

Whey proteins have been employed for microencapsulation to enhance the viability of potential probiotic microorganism by using high internal phase emulsions stabilized with WPI microgels [188,189]. Whey protein hydrogels have shown improved survivability of probiotic bacteria under heat treatment at various storage conditions and along GIT passage [190]. Furthermore, numerous kinds of bio-compounds have also been encapsulated, like tryptophan, riboflavin, and peptides from whey protein microbeads [191], vitamins [192], essential oils [193], curcumin [194], lactoferrin [195], and nutraceuticals like folic acid [196]. WPI nanoparticles have been used for α-tocopherol and resveratrol encapsulation as protein-based carriers for hydrophobic components [197]. WPHs have also been successfully applied for encapsulation of water-soluble nutraceuticals [198].

Incorporation of micro- and nanoparticles as carriers of bioactive compounds entails the controlled delivery of these compounds, thus improving nutritional aspects of functional foods. In addition, enhancement of anticancer activity has been also lately reported as a result of controlled release of lycopene loaded in WPI nanoparticles [199]. Actually, various whey-based matrices have been studied in nutritional applications, incorporating different kind of bioactive compounds [191,200], while the use of whey proteins together with fermentable dietary fibers (such as k-carrageenan), has been also recently reported as a suitable vehicle for the inclusion of proteins and peptides in gelled food products [184].

Likewise, production of whey protein nanofibrils, as novel nanocarriers to enhance the solubility of bioactive compounds and control their release into GIT, constitutes an emerging challenge. Within this context, Alavi et al. [201] employed k-carrageenan, in combination with whey protein aggregates (WPA), in order to control the release of curcumin in the upper gastrointestinal tract [201]. In a similar study, Mohammadian et al. [202] observed the high ability of whey protein nanofibrils to bind curcumin, resulting in a significant release in simulated gastric and intestinal fluids [202]. The authors suggested that whey protein nanofibrils could be used in the formulation of food, drinks, and beverages as a multifunctional carrier for bioactive compounds. The produced hydrogels protected curcumin and proved to be effective for colon-specific delivery. Besides the approach of whey protein gel, proteins can independently assemble to form fibrillary systems. These nanosystems convey new insight into food science applications, exhibiting important functional characteristics, including emulsification and gelation properties, increased viscosity competence, and foam stabilization properties at relatively low protein concentration [203–205].

On top of that, the fabrication of various types of nutraceutical-carrying nanosystems has recently attracted special interest. Nutraceuticals delivery, to beyond water-soluble compounds, could be achieved by the development of novel whey protein hydrogels. In the frame of that, Hashemi et al. [206] proposed the development of gels, through combination of whey protein solution with nanostructured lipid carriers (NLCs) of fat-soluble compounds [206]. Likewise, whey proteins have been described as effective carriers of lipophilic nutraceuticals and scientific attention has been ascribed to the formation of whey-derived products with inhibitory activity on lipid peroxidation. Zhu et al. [207] used surface hydrophobicity properties to form nanocomplexes consisting of whey proteins and fucoxanthin [207]. Prevention of oxidative degradation of carotenoids or other phenolic compounds has been reported, after the inclusion of the aforementioned compounds into WPI emulsions [208–210]. Moreover, increased bioavailability of astaxanthin in Caco-2cell models on whey protein nanodispersion was recently demonstrated [211].

Whey protein nanostructures represent a promising area of food research following their definition as GRAS materials [212]. WPI exhibit core-shell structures similar to natural biopolymers, able to entrap hydrophobic compounds to a great extent. Nanostructural delivery systems are considered substantial approaches to improve biological performance of bioactive compounds. Thus, novel processing techniques are recently employed to modify structural (physicochemicals) and functional properties, demonstrating significant potential for food manufacture applications.

Likewise, pulsed electric field [213], ultrasound [214], or ultraviolet radiation [215] as non-thermal approaches have been shown to exhibit little or no change on the nutritional content. Nevertheless, the electrospinning technique is currently being studied as an emerging advancement for the production of food-grade nanofibers from WP [216]. Electrospinning has been employed to produce micro- to nano-scale fibers as a carrier system to evaluate their potential utilization in food [217]. The ability to generate nanofibers from whey proteins exhibits an opportunity to exploit their inherent benefits, along with the desirable attributes of nanofibers. In addition, incorporation of active compounds and their controlled and monitored subsequent release can be also also achieved [218]. Recent reports dealing with novel electrospun fibers from different blends of WPI have highlighted the advantages of the process, as well as their potential uses in many food related applications [219]. The challenges associated with the development of specific protein fibers correlate with the subsequent specific application, considering that fiber stability in aqueous media and mechanical strength, constitute the most frequent impediments to be challenged [220].

3.3. Whey Protein as a Source of Nutraceuticals

3.3.1. Nutritional Aspects of Whey Proteins

Whey proteins are considered functional nutraceuticals, since they exert remarkable biological activities. They mainly consist of lactoglobulin, lactalbumin, bovine serum albumin, lactoperoxidase, lactoferrin, glycomacropeptide (GMP), and immunoglobulins [6]. From a nutritional aspect, whey proteins are superior to other proteins, such as caseins, since their amino acid profile includes a high proportion of essential branched-chain amino acids (BCAAs) [221,222], such as leucine, isoleucine, and valine, which are crucial in blood glucose homeostasis, metabolism, and neural function [223–225].

Recently, a Leucyl-Valine peptide in whey protein hydrolysate was found to be able to stimulate heat shock proteins response in rats [226]. Heat shock proteins are known to participate in stabilization and restoration of damaged proteins induced by various stress, resulting in maintaining normal cellular function [227]. Whey proteins also contain significant amount of sulfur amino acids, such as methionine and cysteine, which are reported to act as nutraceuticals [228–230]. In general, cheese whey proteins are well-documented as an important source of essential amino acids by means of biologically active peptides. These peptides are considered to be inactive within the sequence of the parent protein, but can be released from whey proteins in sufficient quantities under specific procedures.

Approximately 50% of whey protein is beta-lactoglobulin. Results from a recent study indicate that the Se-β-LG complex presents antitumor activity [231]. The authors also used β- LG nanoparticles as nutraceutical carriers to elevate transepithelial permeation, mucoadhesion, and cellular uptake. In parallel, non-covalent interactions between β-LG and polyphenol extracts of teas, coffee, and cocoa have been reported at pH values of the GIT [232]. The health-promoting effect of cheese whey protein is primarily attributed to their antioxidant properties [233,234] and to their promoting effect on cellular antioxidant pathways [16]. Furthermore, modified products of whey peptides have been shown to increase the antioxidant capacity of the plasma, reducing the risk of certain heart diseases [235]. BSA is another important whey protein, with several drug binding sites, that has been applied as a matrix for nanoparticle-based drug delivery [236].

A broad range of physiological, medical, and nutritional values have been assigned to whey protein and its derivatives, as an excellent source of bioactive peptides. These biomolecules are defined as specific protein fragments that positively influence health and have a beneficial impact on body functions, as summarized in Table 4. Bioactive whey components have been widely studied, revealing various capacities to modulate adiposity, cardiovascular, and gastrointestinal systems [237,238].

Table 4. Whey protein edible film formation from whey protein isolate (WPI) and whey protein isolate (WPI) and their functional features.

Biological Function	Formulation	Test Model	References
Anti-diabetic	Whey protein hydrolysate	Insulin-resistant rats	[239]
	Whey protein	Human	[240]
Anti-inflammatory	β-lactoglobulin hydrolysate	In vitro	[241]
Anti-hypertensive	Whey protein concentrate	In vitro	[242]
Anti-obesity	Whey protein concentrate	Obese human	[243]
	Whey protein concentrate	Obese human	[244]
Antitumor	β-lactoglobulin hydrolysate	*In vitro*	[231]
Benefit in resistant exercise	Hydrolyzed whey protein	Human	[245]
Blood pressure lowering	Whey protein hydrolysate	Rats	[246]
Dermatoprotective	Whey peptide	Mice	[247]
GI motility	Whey protein concentrate, Whey protein Hydrolysate	Mice	[248]
Gut and energy homeostasis	Whey protein isolate	Mice	[249]
Hypolipidemic	Whey protein	Mice	[234]
Muscle protein synthesis/glycogen content	Whey protein hydrolysate	Mice	[226]
Osteroprotection	Whey protein derived dipeptide Glu-Glu	In vitro	[250]
Oxidative stress	Whey protein concentrate	Mice	[251]
Oxidative stress/Glucose metabolism	Whey protein isolate	Overweight/obese patients	[252]
Phenylketonouria therapy	Whey protein glycomacropeptide	Human/mice	[253]
Recovery of muscle functions	Whey protein hydrolysate	Human	[254]
	Whey protein	Human	[255]
Sceletical muscle protection	Whey protein hydrolysate	Rats	[256]

3.3.2. Generation of Bioactive Peptides

Bioactive peptides have attracted significant research and consumer interest, resulting from the potential application in the fortification of products marketed as functional foods or other products for dietary interventions. They can be produced by the following ways: Enzymatic hydrolysis by digestive enzymes during gastrointestinal transit, proteolytic activity of starter cultures during milk fermentation, and proteolytic activity of microorganisms or plants (Figure 3) [257–259]. Microbial fermentation remains an easy and cheap way for generating diverse biopeptides through a safe microbial proteolytic food system. Lately, studies have reported that enzymatic hydrolysis of cheese whey proteins results in different biomolecules with antioxidant properties [260], anticancer [261], and even opioid functions [251]. Alvarado et al. [242] studied the production of antihypertensive peptides from whey protein hydrolysate (<3 kDa) and encapsulated them in order to evaluate angiotensin-converting enzyme activity (ACE%) during GIT digestion [242]. The results revealed about a 10% increase of ACE activity by the released peptides.

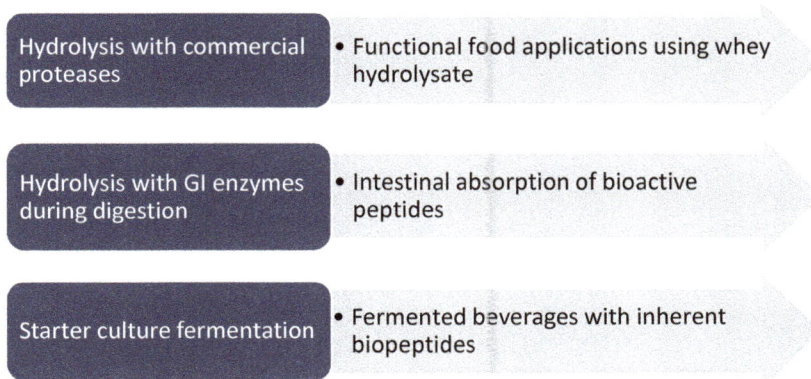

Figure 3. Production methods of bioactive peptides derived from whey proteins and their utilization potential.

All three major forms of cheese whey protein and derivatives (concentrates, isolates, and hydrolysates) encompass unique attributes for nutritional, biological, and food ingredient applications. Recent studies have reported that hydrolysates can maximize nutrient delivery to muscle protein anabolism, presenting higher bioactivity [262–264]. WPHs have been reported for insulinotropic effects [265]. As stated above, many biopeptides are encrypted within their native protein sequences and can thus be liberated only by protein fragmentation [266].

As far as it concerns the industrial production of whey pure protein fractions, techniques of membrane filtration, such as microfiltration and/or ultrafiltration, are employed to enrich whey food ingredients (i.e., whey hydrolysates) [267,268]. Herein, the challenge remains in terms of stability of these fractions under different downstream processes and gastrointestinal phases.

Regardless the scientific achievements, application of bioactive peptides from whey that exert beneficial effect in human nutrition is in its infancy, as several challenges exist in the discovery and identification of bioactive peptide both in vitro and in vivo [269]. Novel bioprocessing strategies for bioactive peptides have been developed as food peptidomics, food proteomics, and nutrigenomic [270,271]. Evidently, an omics approach and bioinformatics could elucidate further development on the application of bioactive peptides [272].

The versatile end applications of whey proteins render them compounds of paramount importance, to be further employed within the frame of integrated biorefineries concepts. However, further research is required to elucidate the underlying mechanisms on food fortification and in human nutrition.

Nonetheless, sedimentation and recovery of protein fraction, separation of individual whey proteins and derivatives synthesis have great potential to be incorporated in an integrated process and further enhance feasibility by obtaining end-products of high-value.

4. Current Integrated Biorefineries

As per the IEA Bioenergy Task 42 definition, "Biorefinery is the sustainable processing of biomass into a spectrum of marketable products (food, feed, materials, chemicals) and energy (fuels, power, heat)" [273]. Thus, to ensure the sustainability and cost-effectiveness of a biorefinery, it is indisputable that a multitude of viable end-products should be manufactured. The development of a biorefinery concept should also employ all potential streams of the onset feedstock under the concept of circular economy and zero waste generation, thus encompassing all three pillars of sustainability, i.e., environment, society, and economy [273]. Integration of biorefining processes annexed to existing manufacturing plants for on-site valorization would alleviate the industry and stake holders' concerns about investing in facilities and equipment that would be depreciated. Another key parameter during the configuration of biorefinery concepts is the complex and heterogenous nature of renewable resources, particularly food waste. The complexity of food waste by-products also impairs the economic assessment of such biorefineries. In any case, high value-added bio-based products deriving from food waste and agricultural by-products should be generated in large amounts, whereas the final market price should be competitive with the chemically produced counterparts [274].

The co-production of fuels and platform chemicals has been the main driver for the configuration of biorefinery scenarios. Biorefining of food waste implements bioprocesses like acidogenesis, fermentation, solventogenesis, oleaginous processes, etc., that yield several products like biofertilizers, animal feed, and biochemicals [14,275]. Nonetheless, bioenergy production is a constructive process for biomass; therefore, to maintain the sustainability, it is advocated that a wide range of bio-products are obtained, instead of a single line product. Several researchers have highlighted the importance of recycling agricultural and industrial wastes through biotransformation by applying a biorefinery concept, utilizing waste as the main feedstock [276].

As previously stated, cheese whey is found among the most significant (and unavoidable) industrial waste streams. Proper disposal and reuse strategies, through bioprocess integration, are unequivocal to mitigate the vast amounts of non-avoidable food waste. Currently, conventional treatments for cheese whey include landfill disposal or anaerobic digestion with focus on BOD and COD reduction, rather than the production of biochemicals, bioenergy, and other value-added novel products [277]. Thus, new approaches for refining have been applied to convert dairy by-products into several valuable bio-based products, such as feed additives, bioplastics, and biochemicals—but also generate high-volume yet lower-value products (bioethanol), or high-value but low-volume products, as nutraceuticals [278].

As it was reported in the previous sections, approaches for cheese whey valorization so far do not exploit the full potential, particularly with respect to food applications. From one point, the protein fraction of cheese whey has been employed to obtain WP to manufacture edible films and coatings, hydrogels, and nutraceuticals (bioactive peptides), among others. The enhanced nutritional value, along with the health benefits, that whey proteins confer [238] have established whey as a high-value raw material in the food industry, with emerging and novel applications. On the other hand, lactose is a precursor substrate for the fermentative synthesis of lactic acid, succinic acid, and polyhydroxyalkanoates, among others [14]. Lactose can be directly fermented by microorganisms, e.g., *Lactobacillus casei*, *Lactobacillus acidophilus*, *Lactobacillus delbrueckii*, *Lactobacillus plantarum*, and *Lactobacillus rhamnosus* [27,279–283]. For instance, lactic acid is an organic biodegradable acid, extensively used in the pharmaceutical, textile, and food industries, where it has been employed in polymerization reactions to produce polylactic acid as a biodegradable polymer. Moreover, lactose deriving from cheese whey can provide a nutrient feedstock for the fermentative production of polyhydroxyalkanoates [284], demonstrating specific physical and mechanical properties.

Biorefining has been already applied for the bioconversion of lactose from cheese whey into several valuable bio-products [10,285–287]; still, the majority of previously reported studies do not implement the complete capacity of cheese whey for food-based formulations. However, within the concept of integrated and consolidated bioprocesses, all streams should be exploited to yield multiple end-products. Likewise, by-products from whey protein manufacture are whey permeate and, following the extraction of lactose, delactosed whey permeate. These dairy-processing side-streams lack effective disposal or further exploitation routes; thus, they can be implemented to formulate cost competitive bio-based products under the concept of sustainable bio-economy targeting "zero waste" [288,289]. Equally, another possible alternative would be to combine different side streams, deriving from separate food processing industries, to configure bioprocesses for multiple end-products.

For instance, a recent study reported the fermentative production of bacterial cellulose (BC) using side-streams of Corinthian currants finishing (CFS), with high antioxidants and sugar content, via the evaluation of nitrogen sources addition and cheese whey [290]. Response surface methodology was applied to evaluate the conditions for BC production for CFS/cheese whey mixtures, concluding that optimum results were achieved on 50.4% whey permeate and 1.7% yeast, at pH 6.36. On top of that, texture analysis was also conducted, indicating that BC could be implemented to formulate foods with potential prebiotic effects, thereby enhancing the functionality of the end-product. Within the same concept, wine lees were combined with cheese whey to develop a biorefining process with microbial oil as the target product [291]. The principal element to initiate process design was the utilization of carbon source from cheese whey (lactose) and the nitrogen source from wine lees, to substitute conventional and expensive chemicals. Polyphenol-rich extracts and tartaric salts, along with crude enzymes via solid state fermentation, were obtained via the treatment of wine lees to yield a nutrient-rich fermentation feedstock. Whey protein concentrate (WPC) was generated after membrane filtration to recover lactose, and at the end of the fermentation process, yeast cell mass could be used as animal feed. The proposed process induced the production of several end-products that could find diversified applications based on market demand, particularly in food formulation (antioxidants, microbial oil, whey protein). It is also worth noting that all streams were exploited, leading to minimal waste generation and enhanced economic feasibility.

A dairy waste biorefinery considering the treatment of cheese whey and cattle manure was proposed and presented by Chandra et al. [28]. Briefly, manure was directed to anaerobic digestion to yield volatile fatty acids, biomethane, hydrogen and fertilizers, considering that biofuel production is often included during a biorefinery design. On the other hand, cheese whey was employed in a more complete valorization scheme to obtain various products. In line with this, lactose was either used for GOS synthesis or for the fermentative production of lactic acid. Alternatively, an enzymatic hydrolysis step for lactose was suggested, prior to alcoholic fermentation or anaerobic digestion, to formulate ethanol. The proposed bioprocess entailed increased viability and sustainability, suggesting also almost zero waste; however, it could be further developed to a configuration that would separately exploit whey protein and lactose into more targeted high value-added food applications, as will be discussed in the following section.

Within the concept of targeted and novel food applications, whey was employed to produce a novel dried cheese whey using thermally-dried *L. casei* ATCC 393 and *L. delbrueckii* ssp. *bulgaricus* ATCC 11842, which were immobilized on casein [292]. Sensory analysis was also performed, and the results clearly demonstrated a novel probiotic product with enhanced aroma, improved shelf life, and protection from pathogenic strains. Hence, to induce novel value-added compounds, it is vital to develop a cheese whey valorization process that is directed principally on products that can separately stand as functional foods or components that are used in food formulation with enhanced properties.

5. Innovative Refining Processes of Cheese Whey and Future Perspectives in Food Applications

Cheese whey is an abundant and low-cost renewable resource deriving from the cheese industry. Utilization of food-grade enzymes and microbial cells, or chemical modification, indicate economically viable approaches for the conversion of cheese whey to produce functional food products. The growing worldwide demand for added-value food products, exhibiting functional properties, is accompanied by an equally increasing market of the latter products to meet consumers' demands. Meanwhile, the transition to bio-economy era demands the sustainable production of these foods, thereby impeding the development of effective bioprocessing and integrated strategies.

The current article demonstrates state-of-the-art processes for cheese whey valorization, towards the manufacture of emerging food additives and products. The majority of the configured processes are focused either only in protein or lactose streams from cheese whey. Evidently, published studies on integrated biorefineries using cheese whey are inadequate with respect to the investigation of both lactose and protein streams to yield food-based compounds. In the era of circular economy, the implementation of several processing methods should be indispensably included within a consolidated cheese whey biorefinery to generate functional foods with improved properties (Figure 4). The following suggested examples relating to food applications indicate solid perspective of novel integrated biorefinery approaches for high value-added food production.

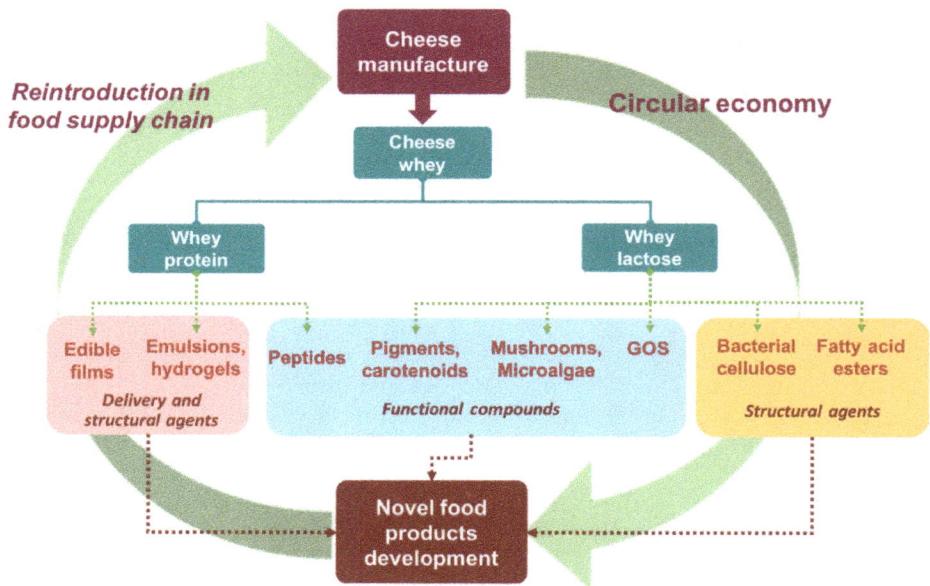

Figure 4. Proposed cheese whey-integrated biorefineries targeting food applications within the circular economy context.

An interesting paradigm to be potentially integrated for cheese whey valorization was introduced by Paximada et al. [293], where whey protein emulsions using bacterial cellulose were produced as an alternative to commercial thickeners, such as xanthan gum and locust bean gum [293]. Xanthan gum is considered an important food thickener, providing high shear-thinning behavior in food products [294]. Worth noting is that Paximada et al. [293] stated that bacterial cellulose had a better shear thinning profile than xanthan gum [293]. Results indicated that lower bacterial cellulose concentration was required compared to xanthan gum or locust bean gum to obtain emulsions with similar rheological properties, and also that bacterial cellulose produced from food by-products is a cheaper alternative to commercial gums [293]. The interaction between whey protein and bacterial cellulose was investigated by Peng et al. [295], noting that bacterial cellulose modified the properties of whey protein fibrillar gel [295]. Particularly, bacterial cellulose addition resulted in a bifibrillar gel, improving whey protein fibril alignment rather than the absence of bacterial cellulose [295].

In Section 2.2.3, we demonstrated that the fermentation of *Spirulina* in cheese whey is an unexploited bioprocess, regardless of the several attempts performed to incorporate the highly nutritious *Spirulina* in foods [134]. Particularly, *Spirulina* has been used as a healthy additive for novel ice cream production with high nutritional value [296]. The estimated cost of the final product was found to be cost competitive within the functional products segment [296]. The addition of *Spirulina platensis* in soft cheese significantly enriched the protein and carotenoid content of the final product [297]. Furthermore, the protein concentrate extracted from the biomass of *Spirulina* sp. has been utilized as functional coating material to encapsulate pigments of commercial interest, such as phycocyanin. Although *Spirulina* sp. is a rich source of this pigment, the production of ultrafine fibers by the electrospinning method resulted in increased thermal stability of phycocyanin [298]. *Spirulina* sp. and *Chlorella* sp. have also been employed for the production of many food products, such as gels [299,300] and fermented milk [135]. In the production of functional fermented milk products, the co-addition of these microalgae and probiotics increased the viability of the probiotic bacteria [135]. Additionally, Terpou et al. [301] reported that whey protein hydrolysate and whey protein concentrate can also promote probiotic viability [301]. These studies indicate that the utilization of both whey protein products and microalgae, previously grown on whey lactose, could be employed for the development of probiotic products with enhanced beneficial value.

This article showed that several protein sources, including *Spirulina* and whey protein, could act as delivery agents for bioactive compounds, including pigments, carotenoids, or GOS, which are produced via microbial or enzymatic bioprocesses using lactose from whey. Additionally, compounds such as lactose esters and bacterial cellulose can be utilized as thickeners and emulsifiers, altering the rheological behavior of food [66,293].

Therefore, to conform to the concept of circular economy that would ideally allow the reintroduction of produced bio-based food components in the food chain, there is a necessity to undertake novel approaches for biorefinery development, considering also the founding pillars of economy, society, and environment. In a complete cascading process, cheese whey would be initially treated to obtain the protein rich fraction and whey lactose. Whey protein fraction could be treated to result in the formulation of nutrient supplements, encapsulation agents of the fermentative production of *Spirulina*, and further inclusion in food. Equally, whey protein could be employed to encapsulate probiotic strains, to manufacture end-products with enhanced nutritional value and sensory characteristics. On the other hand, the lactose-rich stream could be efficiently valorized via enzymatic and microbial bioconversion processes. Synthesis of GOS, either with crude or commercial enzymes, confers a notable option, considering the prominent increase in the market of prebiotics. Purified GOS can be obtained via conventional methods (activated charcoal, membranes, etc.) or via the use of yeast strains (*Kluyveromyces marxianus*, *Saccharomyces cerevisiae*) that consume the unreacted lactose, glucose, and galactose. Appropriate selection of strains in the latter case could lead to additional value-added product formation, whereas the yeast cells after GOS recovery could serve as potential animal feed supplements. The suggested scenario resembles the dairy waste biorefinery proposed by Chandra et al. [28]; however, we elaborate more on high value-added products that will find end applications in the food industry rather than biofuels [28].

Within this concept, our research group at the Department of Food Science & Technology of the Ionian University is currently focusing on the development of an integrated cheese whey biorefinery scheme. More specifically, lactose deriving from whey will be used for the production of potential probiotic starter cultures from non-dairy *Lactobacillus* strains and the production of bacterial cellulose. Subsequently, bacterial cellulose will be combined with whey protein to form immobilization support matrices for probiotic cultures that will be further incorporated in dairy products (e.g., cheese, yogurt). Overall, it is envisaged to configure a refining process to implement both lactose and protein streams, resulting in high value-added products that will be introduced in the food manufacturing sector

(Figure 4). It is easily deduced from these studies that intensive and contemplated effort is conducted to establish an integral process to appraise a closed-loop food supply chain through the manufacture of novel food products. Ultimately, novel approaches will yield alternative bio-based components exhibiting enhanced physicochemical properties, sensory characteristics, and nutritional value.

Author Contributions: Conceptualization, N.K.; investigation, I.K.L., A.P., A.T., V.K., and N.K.; resources, I.K.L., A.P., A.T., and V.K.; writing—original draft preparation, I.K.L., A.P., A.T., V.K., and D.K.; writing—review and editing, A.P., V.K., E.E., and N.K.; supervision, E.E. and N.K.; project administration, N.K.

Funding: This study is part of the project "Valorization of cheese dairy wastes for the production of high added-value products" (MIS 5007020) which is implemented under the Action "Targeted Actions to Promote Research and Technology in Areas of Regional Specialization and New Competitive Areas in International Level", funded by the Operational Programme "Ionian Islands 2014-2020" and co-financed by Greece and the European Union (European Regional Development Fund).

Conflicts of Interest: The authors declare no conflict of interest.

References

1. Lievore, P.; Simões, D.R.S.; Silva, K.M.; Drunkler, N.L.; Barana, A.C.; Nogueira, A.; Demiate, I.M. Chemical characterisation and application of acid whey in fermented milk. *J. Food Sci. Technol.* **2015**, *52*, 2083–2092. [CrossRef] [PubMed]

2. Masotti, F.; Cattaneo, S.; Stuknytė, M.; De Noni, I. Technological tools to include whey proteins in cheese: Current status and perspectives. *Trends Food Sci. Technol.* **2017**, *64*, 102–114. [CrossRef]

3. Fernández-Gutiérrez, D.; Veillette, M.; Giroir-Fendler, A.; Ramirez, A.A.; Faucheux, N.; Heitz, M. Biovalorization of saccharides derived from industrial wastes such as whey: A review. *Rev. Environ. Sci. BioTechnol.* **2017**, *16*, 147–174. [CrossRef]

4. Carvalho, F.; Prazeres, A.R.; Rivas, J. Cheese whey wastewater: Characterization and treatment. *Sci. Total Environ.* **2013**, *445-446*, 385–396. [CrossRef] [PubMed]

5. Wang, Z.; Wang, Z.; Lin, S.; Jin, H.; Gao, S.; Zhu, Y.; Jin, J. Nanoparticle-templated nanofiltration membranes for ultrahigh performance desalination. *Nat. Commun.* **2018**, *9*, 2004. [CrossRef]

6. Yadav, J.S.S.; Yan, S.; Pilli, S.; Kumar, L.; Tyagi, R.D.; Surampalli, R.Y. Cheese whey: A potential resource to transform into bioprotein, functional/nutritional proteins and bioactive peptides. *Biotechnol. Adv.* **2015**, *33*, 756–774. [CrossRef] [PubMed]

7. Chatzipaschali, A.A.; Stamatis, A.G. Biotechnological Utilization with a Focus on Anaerobic Treatment of Cheese Whey: Current Status and Prospects. *Energies* **2012**, *5*, 3492–3525. [CrossRef]

8. Domingos, J.M.B.; Puccio, S.; Martinez, G.A.; Amaral, N.; Reis, M.A.M.; Bandini, S.; Fava, F.; Bertin, L. Cheese whey integrated valorisation: Production, concentration and exploitation of carboxylic acids for the production of polyhydroxyalkanoates by a fed-batch culture. *Chem. Eng. J.* **2018**, *336*, 47–53. [CrossRef]

9. Valta, K.; Damala, P.; Angeli, E.; Antonopoulou, G.; Malamis, D.; Haralambous, K.J. Current Treatment Technologies of Cheese Whey and Wastewater by Greek Cheese Manufacturing Units and Potential Valorisation Opportunities. *Waste Biomass Valorization* **2017**, *8*, 1649–1663. [CrossRef]

10. Remón, J.; Ruiz, J.; Oliva, M.; García, L.; Arauzo, J. Cheese whey valorisation: Production of valuable gaseous and liquid chemicals from lactose by aqueous phase reforming. *Energy Convers. Manag.* **2016**, *124*, 453–469. [CrossRef]

11. Vasala, A.; Panula, J.; Neubauer, P. Efficient lactic acid production from high salt containing dairy by-products by *Lactobacillus salivarius* ssp. salicinius with pre-treatment by proteolytic microorganisms. *J. Biotechnol.* **2005**, *117*, 421–431. [CrossRef] [PubMed]

12. Prazeres, A.R.; Carvalho, F.; Rivas, J. Cheese whey management: A review. *J. Environ. Manag.* **2012**, *110*, 48–68. [CrossRef] [PubMed]

13. Banaszewska, A.; Cruijssen, F.; Claassen, G.D.H.; van der Vorst, J.G.A.J. Effect and key factors of byproducts valorization: The case of dairy industry. *J. Dairy Sci.* **2014**, *97*, 1893–1908. [CrossRef] [PubMed]

14. Koutinas, A.A.; Vlysidis, A.; Pleissner, D.; Kopsahelis, N.; Lopez Garcia, I.; Kookos, I.K.; Papanikolaou, S.; Kwan, T.H.; Lin, C.S.K. Valorization of industrial waste and by-product streams via fermentation for the production of chemicals and biopolymers. *Chem. Soc. Rev.* **2014**, *43*, 2587–2627. [CrossRef] [PubMed]

15. Panghal, A.; Patidar, R.; Jaglan, S.; Chhikara, N.; Khatkar, S.K.; Gat, Y.; Sindhu, N. Whey valorization: Current options and future scenario—A critical review. *Nutr. Food Sci.* **2018**, *48*, 520–535. [CrossRef]

16. Corrochano, A.R.; Buckin, V.; Kelly, P.M.; Giblin, L. Invited review: Whey proteins as antioxidants and promoters of cellular antioxidant pathways. *J. Dairy Sci.* **2018**, *101*, 4747–4761. [CrossRef] [PubMed]

17. Królczyk, J.B.; Dawidziuk, T.; Janiszewska-Turak, E.; Sołowiej, B. Use of Whey and Whey Preparations in the Food Industry—A Review. *Pol. J. Food Nutr. Sci.* **2016**, *66*, 157. [CrossRef]

18. Terpou, A.; Bosnea, L.; Kanellaki, M. Effect of Mastic Gum (*Pistacia Lentiscus Via Chia*) as a Probiotic Cell Encapsulation Carrier for Functional Whey Beverage Production. *SCIOL Biomed.* **2017**, *1*, 1–10.

19. Skryplonek, K.; Dmytrów, I.; Mituniewicz-Małek, A. Probiotic fermented beverages based on acid whey. *J. Dairy Sci.* **2019**, *102*, 7773–7780. [CrossRef] [PubMed]

20. Das, M.; Raychaudhuri, A.; Ghosh, S.K. Supply Chain of Bioethanol Production from Whey: A Review. *Procedia Environ. Sci.* **2016**, *35*, 833–846. [CrossRef]

21. Dedenaro, G.; Costa, S.; Rugiero, I.; Pedrini, P.; Tamburini, E. Valorization of Agri-Food Waste via Fermentation: Production of l-lactic Acid as a Building Block for the Synthesis of Biopolymers. *Appl. Sci.* **2016**, *6*, 379. [CrossRef]

22. Ganju, S.; Gogate, P.R. A review on approaches for efficient recovery of whey proteins from dairy industry effluents. *J. Food Eng.* **2017**, *215*, 84–96. [CrossRef]

23. Uduwerella, G.; Chandrapala, J.; Vasiljevic, T. Preconcentration of yoghurt base by ultrafiltration for reduction in acid whey generation during Greek yoghurt manufacturing. *Int. J. Dairy Technol.* **2018**, *71*, 71–80. [CrossRef]

24. Marx, M.; Kulozik, U. Thermal denaturation kinetics of whey proteins in reverse osmosis and nanofiltration sweet whey concentrates. *Int. Dairy J.* **2018**, *85*, 270–279. [CrossRef]

25. Mawson, A.J. Bioconversions for whey utilization and waste abatement. *Bioresour. Technol.* **1994**, *47*, 195–203. [CrossRef]

26. Pescuma, M.; de Valdez, G.F.; Mozzi, F. Whey-derived valuable products obtained by microbial fermentation. *Appl. Microbiol. Biotechnol.* **2015**, *99*, 6183–6196. [CrossRef] [PubMed]

27. Terpou, A.; Gialleli, A.-I.; Bekatorou, A.; Dimitrellou, D.; Ganatsios, V.; Barouni, E.; Koutinas, A.A.; Kanellaki, M. Sour milk production by wheat bran supported probiotic biocatalyst as starter culture. *Food Bioprod. Process.* **2017**, *101*, 184–192. [CrossRef]

28. Chandra, R.; Castillo-Zacarias, C.; Delgado, P.; Parra-Saldívar, R. A biorefinery approach for dairy wastewater treatment and product recovery towards establishing a biorefinery complexity index. *J. Clean. Prod.* **2018**, *183*, 1184–1196. [CrossRef]

29. Villamiel, M.; Montilla, A.; Olano, A.; Corzo, N. Production and Bioactivity of Oligosaccharides Derived from Lactose. In *Food Oligosaccharides: Production, Analysis and Bioactivity*; John Wiley & Sons, Ltd.: Chichester, UK, 2014; pp. 135–167.

30. Nooshkam, M.; Babazadeh, A.; Jooyandeh, H. Lactulose: Properties, techno-functional food applications, and food grade delivery system. *Trends Food Sci. Technol.* **2018**, *80*, 23–34. [CrossRef]

31. Gibson, G.R.; Roberfroid, M.B. Dietary modulation of the human colonic microbiota: Introducing the concept of prebiotics. *J. Nutr.* **1995**, *125*, 1401–1412. [CrossRef]

32. Gibson, G.R.; Hutkins, R.; Sanders, M.E.; Prescott, S.L.; Reimer, R.A.; Salminen, S.J.; Scott, K.; Stanton, C.; Swanson, K.S.; Cani, P.D.; et al. Expert consensus document: The International Scientific Association for Probiotics and Prebiotics (ISAPP) consensus statement on the definition and scope of prebiotics. *Nat. Rev. Gastroenterol. Hepatol.* **2017**, *14*, 491–502. [CrossRef] [PubMed]

33. Loveren, H.v.; Sanz, Y.; Salminen, S. Health Claims in Europe: Probiotics and Prebiotics as Case Examples. *Annu. Rev. Food Sci. Technol.* **2012**, *3*, 247–261. [CrossRef] [PubMed]

34. Moreno, F.J.; Corzo, N.; Montilla, A.; Villamiel, M.; Olano, A. Current state and latest advances in the concept, production and functionality of prebiotic oligosaccharides. *Curr. Opin. Food Sci.* **2017**, *13*, 50–55. [CrossRef]

35. Sangwan, V.; Tomar, S.K.; Ali, B.; Singh, R.R.B.; Singh, A.K. Production of β-galactosidase from *Streptococcus thermophilus* for galactooligosaccharides synthesis. *J. Food Sci. Technol.* **2015**, *52*, 4206–4215. [CrossRef] [PubMed]

36. Hernández, O.; Ruiz-Matute, A.I.; Olano, A.; Moreno, F.J.; Sanz, M.L. Comparison of fractionation techniques to obtain prebiotic galactooligosaccharides. *Int. Dairy J.* **2009**, *19*, 531–536. [CrossRef]

37. Rastall, R.A. Functional oligosaccharides: Application and manufacture. *Annu. Rev. Food Sci. Technol.* **2010**, *1*, 305–339. [CrossRef] [PubMed]

38. Rastall, R.A. Galacto-Oligosaccharides as Prebiotic Food Ingredients. In *Prebiotics: Development & Application*; John Wiley & Sons, Ltd.: Hoboken, NJ, USA, 2012; pp. 101–109.

39. Adroit market research. Available online: https://www.adroitmarketresearch.com/industry-reports/galacto-oligosaccharides-gos-market (accessed on 15 June 2019).

40. Anadón, A.; Martínez-Larrañaga, M.R.; Arés, I.; Martínez, M.A. Chapter 1—Prebiotics and Probiotics: An Assessment of Their Safety and Health Benefits. In *Probiotics, Prebiotics, and Synbiotics*; Watson, R.R., Preedy, V.R., Eds.; Academic Press: Cambridge, MA, USA, 2016; pp. 3–23.

41. Torres, D.P.M.; Gonçalves, M.d.P.F.; Teixeira, J.A.; Rodrigues, L.R. Galacto-Oligosaccharides: Production, Properties, Applications, and Significance as Prebiotics. *Compr. Rev. Food Sci. Food Saf.* **2010**, *9*, 438–454. [CrossRef]

42. Martins, G.N.; Ureta, M.M.; Tymczyszyn, E.E.; Castilho, P.C.; Gomez-Zavaglia, A. Technological Aspects of the Production of Fructo and Galacto-Oligosaccharides. Enzymatic Synthesis and Hydrolysis. *Front. Nutr.* **2019**, *6*, 78. [CrossRef] [PubMed]

43. Aït-Aissa, A.; Aïder, M. Lactulose: Production and use in functional food, medical and pharmaceutical applications. Practical and critical review. *Int. J. Food Sci. Technol.* **2014**, *49*, 1245–1253. [CrossRef]

44. Contesini, F.J.; de Lima, E.A.; Mandelli, F.; Borin, G.P.; Alves, R.F.; Terrasan, C.R.F. Carbohydrate Active Enzymes Applied in the Production of Functional Oligosaccharides. In *Encyclopedia of Food Chemistry*; Melton, L., Shahidi, F., Varelis, P., Eds.; Academic Press: Oxford, UK, 2019; pp. 30–34.

45. Albayrak, N.; Yang, S.T. Production of galacto-oligosaccharides from lactose by *Aspergillus oryzae* beta-galactosidase immobilized on cotton cloth. *Biotechnol. Bioeng.* **2002**, *77*, 8–19. [CrossRef]

46. Guerrero, C.; Vera, C.; Acevedo, F.; Illanes, A. Simultaneous synthesis of mixtures of lactulose and galacto-oligosaccharides and their selective fermentation. *J. Biotechnol.* **2015**, *209*, 31–40. [CrossRef] [PubMed]

47. Wichienchot, S.; Ishak, W.R.B.W. Prebiotics and Dietary Fibers from Food Processing By-Products. In *Food Processing By-Products and Their Utilization*; John Wiley & Sons Ltd.: Hoboken, NJ, USA, 2017; pp. 137–174.

48. Splechtna, B.; Nguyen, T.-H.; Zehetner, R.; Lettner, H.P.; Lorenz, W.; Haltrich, D. Process development for the production of prebiotic galacto-oligosaccharides from lactose using β-galactosidase from *Lactobacillus* sp. *Biotechnol. J.* **2007**, *2*, 480–485. [CrossRef] [PubMed]

49. Das, R.; Sen, D.; Sarkar, A.; Bhattacharyya, S.; Bhattacharjee, C. A Comparative Study on the Production of Galacto-oligosaccharide from Whey Permeate in Recycle Membrane Reactor and in Enzymatic Batch Reactor. *Ind. Eng. Chem. Res.* **2011**, *50*, 806–816. [CrossRef]

50. Díez-Municio, M.; Montilla, A.; Jimeno, M.L.; Corzo, N.; Olano, A.; Moreno, F.J. Synthesis and Characterization of a Potential Prebiotic Trisaccharide from Cheese Whey Permeate and Sucrose by *Leuconostoc mesenteroides* Dextransucrase. *J. Agric. Food Chem.* **2012**, *60*, 1945–1953. [CrossRef] [PubMed]

51. Corzo-Martínez, M.; Luscher, A.; de Las Rivas, B.; Muñoz, R.; Moreno, F.J. Valorization of Cheese and Tofu Whey through Enzymatic Synthesis of Lactosucrose. *PLoS ONE* **2015**, *10*, e0139035. [CrossRef] [PubMed]

52. Foda, M.I.; Lopez-Leiva, M. Continuous production of oligosaccharides from whey using a membrane reactor. *Process Biochem.* **2000**, *35*, 581–587. [CrossRef]

53. Fischer, C.; Kleinschmidt, T. Synthesis of galactooligosaccharides using sweet and acid whey as a substrate. *Int. Dairy J.* **2015**, *48*, 15–22. [CrossRef]

54. Geiger, B.; Nguyen, H.-M.; Wenig, S.; Nguyen, H.A.; Lorenz, C.; Kittl, R.; Mathiesen, G.; Eijsink, V.G.H.; Haltrich, D.; Nguyen, T.-H. From by-product to valuable components: Efficient enzymatic conversion of lactose in whey using β-galactosidase from *Streptococcus thermophilus*. *BioChem. Eng. J.* **2016**, *116*, 45–53. [CrossRef]

55. Scott, F.; Vera, C.; Conejeros, R. Chapter 7—Technical and Economic Analysis of Industrial Production of Lactose-Derived Prebiotics with Focus on Galacto-Oligosaccharides. In *Lactose-Derived Prebiotics*; Illanes, A., Guerrero, C., Vera, C., Wilson, L., Conejeros, R., Scott, F., Eds.; Academic Press: San Diego, CA, USA, 2016; pp. 261–284.

56. Market Research. Available online: https://www.marketresearch.com/MarketsandMarkets-v3719/Sucrose-Esters-Application-Food-Detergents-9762838/ (accessed on 15 June 2019).

57. Staroń, J.; Dąbrowski, J.M.; Cichoń, E.; Guzik, M. Lactose esters: Synthesis and biotechnological applications. *Crit. Rev. Biotechnol.* **2018**, *38*, 245–258. [CrossRef]

58. Papadaki, A.; Cipolatti, E.P.; Aguieiras, E.C.G.; Cerqueira Pinto, M.C.; Kopsahelis, N.; Freire, D.M.G.; Mandala, I.; Koutinas, A.A. Development of Microbial Oil Wax-Based Oleogel with Potential Application in Food Formulations. *Food Bioprocess Technol.* **2019**, *12*, 899–909. [CrossRef]

59. Papadaki, A.; Fernandes, K.V.; Chatzifragkou, A.; Aguieiras, E.C.G.; da Silva, J.A.C.; Fernandez-Lafuente, R.; Papanikolaou, S.; Koutinas, A.; Freire, D.M.G. Bioprocess development for biolubricant production using microbial oil derived via fermentation from confectionery industry wastes. *Bioresour. Technol.* **2018**, *267*, 311–318. [CrossRef] [PubMed]

60. Papadaki, A.; Mallouchos, A.; Efthymiou, M.-N.; Gardeli, C.; Kopsahelis, N.; Aguieiras, E.C.G.; Freire, D.M.G.; Papanikolaou, S.; Koutinas, A.A. Production of wax esters via microbial oil synthesis from food industry waste and by-product streams. *Bioresour. Technol.* **2017**, *245*, 274–282. [CrossRef] [PubMed]

61. Fernandes, K.V.; Papadaki, A.; da Silva, J.A.C.; Fernandez-Lafuente, R.; Koutinas, A.A.; Freire, D.M.G. Enzymatic esterification of palm fatty-acid distillate for the production of polyol esters with biolubricant properties. *Ind. Crop. Prod.* **2018**, *116*, 90–96. [CrossRef]

62. Scholnick, F.; Sucharski, M.K.; Linfield, W.M. Lactose-derived surfactants (I) fatty esters of lactose. *J. Am. Oil Chem. Soc.* **1974**, *51*, 8–11. [CrossRef]

63. Lay, L.; Panza, L.; Riva, S.; Khitri, M.; Tirendi, S. Regioselective acylation of disaccharides by enzymatic transesterification. *Carbohydr. Res.* **1996**, *291*, 197–204. [CrossRef]

64. Riva, S.; Chopineau, J.; Kieboom, A.P.G.; Klibanov, A.M. Protease-catalyzed regioselective esterification of sugars and related compounds in anhydrous dimethylformamide. *J. Am. Chem. Soc.* **1988**, *110*, 584–589. [CrossRef]

65. Enayati, M.; Gong, Y.; Goddard, J.M.; Abbaspourrad, A. Synthesis and characterization of lactose fatty acid ester biosurfactants using free and immobilized lipases in organic solvents. *Food Chem.* **2018**, *266*, 508–513. [CrossRef]

66. Lee, S.-M.; Wagh, A.; Sandhu, G.; Walsh, M.K. Emulsification Properties of Lactose Fatty Acid Esters. *Food Nutr. Sci.* **2018**, *09*, 17. [CrossRef]

67. Ambati, R.R.; Phang, S.M.; Ravi, S.; Aswathanarayana, R.G. Astaxanthin: Sources, extraction, stability, biological activities and its commercial applications—A review. *Mar. Drugs* **2014**, *12*, 128–152. [CrossRef]

68. Tanumihardjo, S.A. Carotenoids: Health Effects. In *Encyclopedia of Human Nutrition*, 3rd ed.; Caballero, B., Ed.; Academic Press: Waltham, MA, USA, 2013; pp. 292–297.

69. Eggersdorfer, M.; Wyss, A. Carotenoids in human nutrition and health. *Arch. Biochem. Biophys.* **2018**, *652*, 18–26. [CrossRef]

70. Bogacz-Radomska, L.; Harasym, J. β-Carotene—Properties and production methods. *Food Qual. Saf.* **2018**, *2*, 69–74. [CrossRef]

71. Mata-Gómez, L.C.; Montañez, J.C.; Méndez-Zavala, A.; Aguilar, C.N. Biotechnological production of carotenoids by yeasts: An overview. *Microb. Cell Fact.* **2014**, *13*, 12. [CrossRef] [PubMed]

72. Shah, M.M.R.; Liang, Y.; Cheng, J.J.; Daroch, M. Astaxanthin-Producing Green Microalga *Haematococcus pluvialis*: From Single Cell to High Value Commercial Products. *Front. Plant Sci.* **2016**, *7*, 531. [CrossRef] [PubMed]

73. Yuan, J.P.; Peng, J.; Yin, K.; Wang, J.H. Potential health-promoting effects of astaxanthin: A high-value carotenoid mostly from microalgae. *Mol. Nutr. Food Res.* **2011**, *55*, 150–165. [CrossRef] [PubMed]

74. Villegas-Méndez, M.Á.; Aguilar-Machado, D.E.; Balagurusamy, N.; Montañez, J.; Morales-Oyervides, L. Agro-industrial wastes for the synthesis of carotenoids by *Xanthophyllomyces dendrorhous*: Mesquite pods-based medium design and optimization. *BioChem. Eng. J.* **2019**, *150*, 107260. [CrossRef]

75. Barredo, J.L.; García-Estrada, C.; Kosalkova, K.; Barreiro, C. Biosynthesis of Astaxanthin as a Main Carotenoid in the Heterobasidiomycetous Yeast *Xanthophyllomyces dendrorhous*. *J. Fungi* **2017**, *3*, 44. [CrossRef]

76. Roukas, T.; Varzakakou, M.; Kotzekidou, P. From Cheese Whey to Carotenes by *Blakeslea trispora* in a Bubble Column Reactor. *Appl. Biochem. Biotechnol.* **2015**, *175*, 182–193. [CrossRef] [PubMed]

77. Khodaiyan, F.; Razavi, S.H.; Mousavi, S.M. Optimization of canthaxanthin production by *Dietzia natronolimnaea* HS-1 from cheese whey using statistical experimental methods. *BioChem. Eng. J.* **2008**, *40*, 415–422. [CrossRef]

78. Liu, Z.-Q.; Zhang, J.-F.; Zheng, Y.-G.; Shen, Y.-C. Improvement of astaxanthin production by a newly isolated *Phaffia rhodozyma* mutant with low-energy ion beam implantation. *J. Appl. Microbiol.* **2008**, *104*, 861–872. [CrossRef]

79. Sun, N.; Wang, Y.; Li, Y.-T.; Huang, J.-C.; Chen, F. Sugar-based growth, astaxanthin accumulation and carotenogenic transcription of heterotrophic *Chlorella zofingiensis* (Chlorophyta). *Process Biochem.* **2008**, *43*, 1288–1292. [CrossRef]

80. Aksu, Z.; Eren, A.T. Carotenoids production by the yeast Rhodotorula mucilaginosa: Use of agricultural wastes as a carbon source. *Process Biochem.* **2005**, *40*, 2985–2991. [CrossRef]

81. Mantzouridou, F.; Tsimidou, M.Z.; Roukas, T. Performance of Crude Olive Pomace Oil and Soybean Oil during Carotenoid Production by Blakeslea trispora in Submerged Fermentation. *J. Agric. Food Chem.* **2006**, *54*, 2575–2581. [CrossRef] [PubMed]

82. Varzakakou, M.; Roukas, T. Identification of carotenoids produced from cheese whey by *Blakeslea trispora* in submerged fermentation. *Prep. Biochem. Biotechnol.* **2009**, *40*, 76–82. [CrossRef] [PubMed]

83. Varzakakou, M.; Roukas, T.; Kotzekidou, P. Effect of the ratio of (+) and (−) mating type of *Blakeslea trispora* on carotene production from cheese whey in submerged fermentation. *World J. Microbiol. Biotechnol.* **2010**, *26*, 2151–2156. [CrossRef]

84. Roukas, T.; Mantzouridou, F.; Boumpa, T.; Vafiadou, A.; Goksungur, Y. Production of β-Carotene from Beet Molasses and Deproteinized Whey by *Blakeslea trispora*. *Food Biotechnol.* **2007**, *17*, 203–215. [CrossRef]

85. Psani, M.; Roukas, T.; Kotzekidou, P. Evaluation of cheese whey as substrate for carotenoids production by *Blakeslea trispora*. *Australian J. Dairy Technol.* **2006**, *61*, 222.

86. Azmi, W.; Thakur, M.; Kumar, A. Production of beta-carotene from deproteinized waste whey filtrate using Mucor azygosporus MTCC 414 in submerged fermentation. *Acta Microbiol. Immunol. Hung.* **2011**, *58*, 189–200. [CrossRef] [PubMed]

87. Marova, I.; Carnecka, M.; Halienova, A.; Certik, M.; Dvorakova, T.; Haronikova, A. Use of several waste substrates for carotenoid-rich yeast biomass production. *J. Environ. Manag.* **2012**, *95*, S338–S342. [CrossRef]

88. Frengova, G.; Simova, E.; Beshkova, D. Use of whey ultrafiltrate as a substrate for production of carotenoids by the yeast *Rhodotorula rubra*. *Appl. Biochem. Biotechnol.* **2004**, *112*, 133–141. [CrossRef]

89. Frengova, G.; Simova, E.; Pavlova, K.; Beshkova, D.; Grigorova, D. Formation of carotenoids by *Rhodotorula glutinis* in whey ultrafiltrate. *Biotechnol. Bioeng.* **1994**, *44*, 888–894. [CrossRef]

90. Simova, E.D.; Frengova, G.I.; Beshkova, D.M. Synthesis of carotenoids by *Rhodotorula rubra* GED8 co-cultured with yogurt starter cultures in whey ultrafiltrate. *J. Ind. Microbiol. Biotechnol.* **2004**, *31*, 115–121. [CrossRef] [PubMed]

91. Valduga, E.; Tatsch, P.; Vanzo, L.T.; Rauber, F.; Di Luccio, M.; Treichel, H. Assessment of hydrolysis of cheese whey and use of hydrolysate for bioproduction of carotenoids by *Sporidiobolus salmonicolor* CBS 2636. *J. Sci. Food Agric.* **2009**, *89*, 1060–1065. [CrossRef]

92. Li, X.; Wang, D.; Cai, D.; Zhan, Y.; Wang, Q.; Chen, S. Identification and High-level Production of Pulcherrimin in *Bacillus licheniformis* DW2. *Appl. Biochem. Biotechnol.* **2017**, *183*, 1323–1335. [CrossRef] [PubMed]

93. Turkel, S.; Korukluoglu, M.; Yavuz, M. Biocontrol Activity of the Local Strain of *Metschnikowia pulcherrima* on Different Postharvest Pathogens. *Biotechnol. Res. Int.* **2014**, *2014*, 397167. [CrossRef]

94. Türkel, S.; Ener, B. Isolation and Characterization of New *Metschnikowia pulcherrima* Strains as Producers of the Antimicrobial Pigment Pulcherrimin. *Z. Naturforsch. C* **2009**, *64*, 405–410. [CrossRef]

95. Savini, V.; Hendrickx, M.; Sisti, M.; Masciarelli, G.; Favaro, M.; Fontana, C.; Pitzurra, L.; Arzeni, D.; Astolfi, D.; Catavitello, C.; et al. An atypical, pigment-producing Metschnikowia strain from a leukaemia patient. *Med. Mycol.* **2013**, *51*, 438–443. [CrossRef] [PubMed]

96. Chreptowicz, K.; Sternicka, M.K.; Kowalska, P.D.; Mierzejewska, J. Screening of yeasts for the production of 2-phenylethanol (rose aroma) in organic waste-based media. *Lett. Appl. Microbiol.* **2018**, *66*, 153–160. [CrossRef]

97. Chantasuban, T.; Santomauro, F.; Gore-Lloyd, D.; Parsons, S.; Henk, D.; Scott, R.J.; Chuck, C. Elevated production of the aromatic fragrance molecule, 2-phenylethanol, using *Metschnikowia pulcherrima* through both de novo and ex novo conversion in batch and continuous modes. *J. Chem. Technol. Biotechnol.* **2018**, *93*, 2118–2130. [CrossRef]

98. Izawa, N.; Kudo, M.; Nakamura, Y.; Mizukoshi, H.; Kitada, T.; Sone, T. Production of aroma compounds from whey using *Wickerhamomyces pijperi*. *AMB Express* **2015**, *5*, 23. [CrossRef]

99. Azeredo, H.M.C.; Barud, H.; Farinas, C.S.; Vasconcellos, V.M.; Claro, A.M. Bacterial Cellulose as a Raw Material for Food and Food Packaging Applications. *Front. Sustain. Food Syst.* **2019**, *3*. [CrossRef]

100. Tsouko, E.; Kourmentza, C.; Ladakis, D.; Kopsahelis, N.; Mandala, I.; Papanikolaou, S.; Paloukis, F.; Alves, V.; Koutinas, A. Bacterial Cellulose Production from Industrial Waste and by-Product Streams. *Int. J. Mol. Sci.* **2015**, *16*, 14832–14849. [CrossRef] [PubMed]

101. Andritsou, V.; de Melo, E.M.; Tsouko, E.; Ladakis, D.; Maragkoudaki, S.; Koutinas, A.A.; Matharu, A.S. Synthesis and Characterization of Bacterial Cellulose from Citrus-Based Sustainable Resources. *ACS Omega* **2018**, *3*, 10365–10373. [CrossRef]

102. Cacicedo, M.L.; Castro, M.C.; Servetas, I.; Bosnea, L.; Boura, K.; Tsafrakidou, P.; Dima, A.; Terpou, A.; Koutinas, A.; Castro, G.R. Progress in bacterial cellulose matrices for biotechnological applications. *Bioresour. Technol.* **2016**, *213*, 172–180. [CrossRef] [PubMed]

103. Vazquez, A.; Foresti, M.L.; Cerrutti, P.; Galvagno, M. Bacterial Cellulose from Simple and Low Cost Production Media by *Gluconacetobacter xylinus*. *J. Polym. Environ.* **2013**, *21*, 545–554. [CrossRef]

104. Tsouko, E.; Papadaki, A.; Papapostolou, H.; Ladakis, D.; Natsia, A.; Koutinas, A.; Kampioti, A.; Eriotou, E.; Kopsahelis, N. Valorization of Zante currant side-streams for the production of phenolic-rich extract and bacterial cellulose: A novel biorefinery concept. *J. Chem. Technol. Biotechnol.* **2019**. [CrossRef]

105. Lee, K.-Y.; Buldum, G.; Mantalaris, A.; Bismarck, A. More Than Meets the Eye in Bacterial Cellulose: Biosynthesis, Bioprocessing, and Applications in Advanced Fiber Composites. *Macromol. Biosci.* **2014**, *14*, 10–32. [CrossRef] [PubMed]

106. Dahman, Y.; Jayasuriya, K.E.; Kalis, M. Potential of Biocellulose Nanofibers Production from Agricultural Renewable Resources: Preliminary Study. *Appl. Biochem. Biotechnol.* **2010**, *162*, 1647–1659. [CrossRef]

107. Son, H.-J.; Kim, H.-G.; Kim, K.-K.; Kim, H.-S.; Kim, Y.-G.; Lee, S.-J. Increased production of bacterial cellulose by *Acetobacter* sp. V6 in synthetic media under shaking culture conditions. *Bioresour. Technol.* **2003**, *86*, 215–219. [CrossRef]

108. Mikkelsen, D.; Flanagan, B.M.; Dykes, G.A.; Gidley, M.J. Influence of different carbon sources on bacterial cellulose production by *Gluconacetobacter xylinus* strain ATCC 53524. *J. Appl. Microbiol.* **2009**, *107*, 576–583. [CrossRef]

109. Thompson, D.N.; Hamilton, M.A. Production of bacterial cellulose from alternate feedstocks. *Appl. Biochem. Biotechnol.* **2001**, *91*, 503. [CrossRef]

110. Carreira, P.; Mendes, J.A.S.; Trovatti, E.; Serafim, L.S.; Freire, C.S.R.; Silvestre, A.J.D.; Neto, C.P. Utilization of residues from agro-forest industries in the production of high value bacterial cellulose. *Bioresour. Technol.* **2011**, *102*, 7354–7360. [CrossRef] [PubMed]

111. Battad-Bernardo, E.; McCrindle, S.L.; Couperwhite, I.; Neilan, B.A. Insertion of an E. coli lacZ gene in *Acetobacter xylinus* for the production of cellulose in whey. *FEMS Microbiol. Lett.* **2004**, *231*, 253–260. [CrossRef]

112. Salari, M.; Sowti Khiabani, M.; Rezaei Mokarram, R.; Ghanbarzadeh, B.; Samadi Kafil, H. Preparation and characterization of cellulose nanocrystals from bacterial cellulose produced in sugar beet molasses and cheese whey media. *Int. J. Biol. Macromol.* **2019**, *122*, 280–288. [CrossRef] [PubMed]

113. Revin, V.; Liyaskina, E.; Nazarkina, M.; Bogatyreva, A.; Shchankin, M. Cost-effective production of bacterial cellulose using acidic food industry by-products. *Braz. J. Microbiol.* **2018**, *49*, 151–159. [CrossRef] [PubMed]

114. Kothari, D.; Patel, S.; Kim, S.-K. Anticancer and other therapeutic relevance of mushroom polysaccharides: A holistic appraisal. *Biomed. Pharmacother.* **2018**, *105*, 377–394. [CrossRef]

115. Borthakur, M.; Joshi, S.R. Chapter 1—Wild Mushrooms as Functional Foods: The Significance of Inherent Perilous Metabolites. In *New and Future Developments in Microbial Biotechnology Bioengineering*; Gupta, V.K., Pandey, A., Eds.; Elsevier: Amsterdam, The Netherlands, 2019; pp. 1–12.

116. Diamantopoulou, P.; Papanikolaou, S.; Kapoti, M.; Komaitis, M.; Aggelis, G.; Philippoussis, A. Mushroom Polysaccharides and Lipids Synthesized in Liquid Agitated and Static Cultures. Part I: Screening Various Mushroom Species. *Appl. Biochem. Biotechnol.* **2012**, *167*, 536–551. [CrossRef]

117. Hereher, F.; ElFallal, A.; Toson, E.; Abou-Dobara, M.; Abdelaziz, M. Pilot study: Tumor suppressive effect of crude polysaccharide substances extracted from some selected mushroom. *Beni-Suef Univ. J. Basic Appl. Sci.* **2018**, *7*, 767–775. [CrossRef]

118. Giavasis, I. Polysaccharides from Medicinal Mushrooms for Potential Use as Nutraceuticals. In *Polysaccharides Natural Fibers in Food and Nutrition*; Benkeblia, N., Ed.; CRC Press: Boca Raton, FL, USA, 2014.

119. Velez, M.E.V.; da Luz, J.M.R.; da Silva, M.d.C.S.; Cardoso, W.S.; Lopes, L.d.S.; Vieira, N.A.; Kasuya, M.C.M. Production of bioactive compounds by the mycelial growth of *Pleurotus djamor* in whey powder enriched with selenium. *LWT* **2019**, *114*, 108376. [CrossRef]

120. Mukhopadhyay, R.; Guha, A.K. A comprehensive analysis of the nutritional quality of edible mushroom *Pleurotus sajor-caju* grown in deproteinized whey medium. *LWT Food Sci. Technol.* **2015**, *61*, 339–345. [CrossRef]

121. Wu, X.J. Proximate Composition of *Pleurotus ostreatus* Grown in Whey Permeate Based Medium. *Trans. ASABE* **2009**, *52*, 1249–1254.

122. Bhak, G.; Song, M.; Lee, S.; Hwang, S. Response Surface Analysis of Solid State Growth of *Pleurotus ostreatus* Mycelia utilizing Whey Permeate. *Biotechnol. Lett.* **2005**, *27*, 1537–1541. [CrossRef] [PubMed]

123. Israilides, C.; Philippoussis, A. Bio-technologies of Recycling Agro-industrial Wastes for the Production of Commercially Important Fungal Polysaccharides and Mushrooms. *Biotechnol. Genet. Eng. Rev.* **2003**, *20*, 247–260. [CrossRef] [PubMed]

124. Wu, X.J.; Hansen, C. Effects of Whey Permeate-Based Medium on the Proximate Composition of *Lentinus edodes* in the Submerged Culture. *J. Food Sci.* **2006**, *71*, M174–M179. [CrossRef]

125. Wu, X.J.; Hansen, C. Antioxidant Capacity, Phenolic Content, and Polysaccharide Content of Lentinus edodes Grown in Whey Permeate-Based Submerged Culture. *J. Food Sci.* **2008**, *73*, M1–M8. [CrossRef] [PubMed]

126. Inglet, B.S.; Song, M.; Hansen, C.L.; Hwang, S. Short Communication: Cultivation of *Lentinus edodes* Mycelia Using Whey Permeate as an Alternative Growth Substrate. *J. Dairy Sci.* **2006**, *89*, 1113–1115. [CrossRef]

127. Song, M.; Kim, N.; Lee, S.; Hwang, S. Use of Whey Permeate for Cultivating *Ganoderma lucidum* Mycelia. *J. Dairy Sci.* **2007**, *90*, 2141–2146. [CrossRef] [PubMed]

128. Lee, H.; Song, M.; Hwang, S. Optimizing bioconversion of deproteinated cheese whey to mycelia of *Ganoderma lucidum*. *Process Biochem.* **2003**, *38*, 1685–1693. [CrossRef]

129. Lee, H.; Song, M.; Yu, Y.; Hwang, S. Production of *Ganoderma lucidum* mycelium using cheese whey as an alternative substrate: Response surface analysis and biokinetics. *BioChem. Eng. J.* **2003**, *15*, 93–99. [CrossRef]

130. Sanodiya, B.S.; Thakur, G.S.; Baghel, R.K.; Prasad, G.B.; Bisen, P.S. *Ganoderma lucidum*: A potent pharmacological macrofungus. *Curr. Pharm. Biotechnol.* **2009**, *10*, 717–742. [CrossRef]

131. Shao, P.; Xuan, S.; Wu, W.; Qu, L. Encapsulation efficiency and controlled release of *Ganoderma lucidum* polysaccharide microcapsules by spray drying using different combinations of wall materials. *Int. J. Biol. Macromol.* **2019**, *125*, 962–969. [CrossRef]

132. Papadaki, A.; Diamantopoulou, P.; Papanikolaou, S.; Philippoussis, A. Evaluation of Biomass and Chitin Production of *Morchella* Mushrooms Grown on Starch-Based Substrates. *Foods* **2019**, *8*, 239. [CrossRef] [PubMed]

133. Kosaric, N.; Miyata, N. Growth of morel mushroom mycelium in cheese whey. *J. Dairy Res.* **1981**, *48*, 149–162. [CrossRef]

134. Caporgno, M.P.; Mathys, A. Trends in Microalgae Incorporation into Innovative Food Products with Potential Health Benefits. *Front. Nutr.* **2018**, *5*, 58. [CrossRef] [PubMed]

135. Beheshtipour, H.; Mortazavian, A.M.; Mohammadi, R.; Sohrabvandi, S.; Khosravi-Darani, K. Supplementation of *Spirulina platensis* and Chlorella vulgaris Algae into Probiotic Fermented Milks. *Compr. Rev. Food Sci. Food Saf.* **2013**, *12*, 144–154. [CrossRef]

136. Abreu, A.P.; Fernandes, B.; Vicente, A.A.; Teixeira, J.; Dragone, G. Mixotrophic cultivation of *Chlorella vulgaris* using industrial dairy waste as organic carbon source. *Bioresour. Technol.* **2012**, *118*, 61–66. [CrossRef] [PubMed]

137. Kulandaivel, S.; Prakash, R.; Anitha, R.; Arunnagendran, N. Comparative studies on biochemical profile of *Spirulina platensis* and *Oscillatoria sp.* on synthetic medium and dairy effluent. *J. Pure Appl. Microbiol.* **2007**, *1*, 109–112.

138. Vieira Salla, A.C.; Margarites, A.C.; Seibel, F.I.; Holz, L.C.; Brião, V.B.; Bertolin, T.E.; Colla, L.M.; Costa, J.A.V. Increase in the carbohydrate content of the microalgae *Spirulina* in culture by nutrient starvation and the addition of residues of whey protein concentrate. *Bioresour. Technol.* **2016**, *209*, 133–141. [CrossRef] [PubMed]

139. Girard, J.-M.; Roy, M.-L.; Hafsa, M.B.; Gagnon, J.; Faucheux, N.; Heitz, M.; Tremblay, R.; Deschênes, J.-S. Mixotrophic cultivation of green microalgae *Scenedesmus obliquus* on cheese whey permeate for biodiesel production. *Algal Res.* **2014**, *5*, 241–248. [CrossRef]

140. Afify, A.E.-M.M.R.; El Baroty, G.S.; El Baz, F.K.; Abd El Baky, H.H.; Murad, S.A. Scenedesmus obliquus: Antioxidant and antiviral activity of proteins hydrolyzed by three enzymes. *J. Genet. Eng. Biotechnol.* **2018**, *16*, 399–408. [CrossRef] [PubMed]

141. Bleakley, S.; Hayes, M. Algal Proteins: Extraction, Application, and Challenges Concerning Production. *Foods* **2017**, *6*, 33. [CrossRef] [PubMed]

142. Cinelli, P.; Schmid, M.; Bugnicourt, E.; Wildner, J.; Bazzichi, A.; Anguillesi, I.; Lazzeri, A. Whey protein layer applied on biodegradable packaging film to improve barrier properties while maintaining biodegradability. *Polym. Degrad. Stab.* **2014**, *108*, 151–157. [CrossRef]

143. Tarhan, O.; Spotti, M.J.; Schaffter, S.; Corvalan, C.M.; Campanella, O.H. Rheological and structural characterization of whey protein gelation induced by enzymatic hydrolysis. *Food Hydrocoll.* **2016**, *61*, 211–220. [CrossRef]

144. Fu, W.; Nakamura, T. Explaining the texture properties of whey protein isolate/starch co-gels from fracture structures. *Biosci. Biotechnol. Biochem.* **2017**, *81*, 839–847. [CrossRef] [PubMed]

145. Khalifa, I.; Nie, R.; Ge, Z.; Li, K.; Li, C. Understanding the shielding effects of whey protein on mulberry anthocyanins: Insights from multispectral and molecular modelling investigations. *Int. J. Biol. Macromol.* **2018**, *119*, 116–124. [CrossRef] [PubMed]

146. Andoyo, R.; Dianti Lestari, V.; Mardawati, E.; Nurhadi, B. Fractal Dimension Analysis of Texture Formation of Whey Protein-Based Foods. *Int. J. Food Sci.* **2018**, *2018*. [CrossRef] [PubMed]

147. Kurek, M.; Galus, S.; Debeaufort, F. Surface, mechanical and barrier properties of bio-based composite films based on chitosan and whey protein. *Food Packag. Shelf Life* **2014**, *1*, 56–67. [CrossRef]

148. Galus, S.; Kadzińska, J. Whey protein edible films modified with almond and walnut oils. *Food Hydrocoll.* **2016**, *52*, 78–86. [CrossRef]

149. Barba, C.; Eguinoa, A.; Maté, J.I. Preparation and characterization of β-cyclodextrin inclusion complexes as a tool of a controlled antimicrobial release in whey protein edible films. *LWT Food Sci. Technol.* **2015**, *64*, 1362–1369. [CrossRef]

150. Boyacı, D.; Korel, F.; Yemenicioğlu, A. Development of activate-at-home-type edible antimicrobial films: An example pH-triggering mechanism formed for smoked salmon slices using lysozyme in whey protein films. *Food Hydrocoll.* **2016**, *60*, 170–178. [CrossRef]

151. Schmid, M.; Merzbacher, S.; Brzoska, N.; Müller, K.; Jesdinszki, M. Improvement of Food Packaging-Related Properties of Whey Protein Isolate-Based Nanocomposite Films and Coatings by Addition of Montmorillonite Nanoplatelets. *Front. Mater.* **2017**, *4*. [CrossRef]

152. Azeredo, H.M.C.; Waldron, K.W. Crosslinking in polysaccharide and protein films and coatings for food contact—A review. *Trends Food Sci. Technol.* **2016**, *52*, 109–122. [CrossRef]

153. Jiang, S.-J.; Zhang, T.; Song, Y.; Qian, F.; Tuo, Y.; Mu, G. Mechanical properties of whey protein concentrate based film improved by the coexistence of nanocrystalline cellulose and transglutaminase. *Int. J. Biol. Macromol.* **2019**, *126*, 1266–1272. [CrossRef] [PubMed]

154. Qazanfarzadeh, Z.; Kadivar, M. Properties of whey protein isolate nanocomposite films reinforced with nanocellulose isolated from oat husk. *Int. J. Biol. Macromol.* **2016**, *91*, 1134–1140. [CrossRef] [PubMed]

155. Hassannia-Kolaee, M.; Khodaiyan, F.; Pourahmad, R.; Shahabi-Ghahfarrokhi, I. Development of ecofriendly bionanocomposite: Whey protein isolate/pullulan films with nano-SiO$_2$. *Int. J. Biol. Macromol.* **2016**, *86*, 139–144. [CrossRef] [PubMed]

156. Zhang, W.; Chen, J.; Chen, Y.; Xia, W.; Xiong, Y.L.; Wang, H. Enhanced physicochemical properties of chitosan/whey protein isolate composite film by sodium laurate-modified TiO$_2$ nanoparticles. *Carbohydr. Polym.* **2016**, *138*, 59–65. [CrossRef]

157. Basiak, E.; Lenart, A.; Debeaufort, F. Effects of carbohydrate/protein ratio on the microstructure and the barrier and sorption properties of wheat starch–whey protein blend edible films. *J. Sci. Food Agric.* **2017**, *97*, 858–867. [CrossRef] [PubMed]

158. Tsai, M.-J.; Weng, Y.-M. Novel edible composite films fabricated with whey protein isolate and zein: Preparation and physicochemical property evaluation. *LWT* **2019**, *101*, 567–574. [CrossRef]

159. Oymaci, P.; Altinkaya, S.A. Improvement of barrier and mechanical properties of whey protein isolate based food packaging films by incorporation of zein nanoparticles as a novel bionanocomposite. *Food Hydrocoll.* **2016**, *54*, 1–9. [CrossRef]

160. Bahram, S.; Rezaei, M.; Soltani, M.; Kamali, A.; Ojagh, S.M.; Abdollahi, M. Whey Protein Concentrate Edible Film Activated with Cinnamon Essential Oil. *J. Food Process. Preserv.* **2014**, *38*, 1251–1258. [CrossRef]

161. Khanzadi, M.; Jafari, S.M.; Mirzaei, H.; Chegini, F.K.; Maghsoudlou, Y.; Dehnad, D. Physical and mechanical properties in biodegradable films of whey protein concentrate–pullulan by application of beeswax. *Carbohydr. Polym.* **2015**, *118*, 24–29. [CrossRef]

162. Rantamäki, P.; Loimaranta, V.; Vasara, E.; Latva-Koivisto, J.; Korhonen, H.; Tenovuo, J.; Marnila, P. Edible films based on milk proteins release effectively active immunoglobulins. *Food Qual. Saf.* **2019**, *3*, 23–34. [CrossRef]

163. Piccirilli, G.N.; Soazo, M.; Pérez, L.M.; Delorenzi, N.J.; Verdini, R.A. Effect of storage conditions on the physicochemical characteristics of edible films based on whey protein concentrate and liquid smoke. *Food Hydrocoll.* **2019**, *87*, 221–228. [CrossRef]

164. Pereira, R.C.; de Deus Souza Carneiro, J.; Borges, S.V.; Assis, O.B.G.; Alvarenga, G.L. Preparation and Characterization of Nanocomposites from Whey Protein Concentrate Activated with Lycopene. *J. Food Sci.* **2016**, *81*, E637–E642. [CrossRef] [PubMed]

165. Soukoulis, C.; Behboudi-Jobbehdar, S.; Macnaughtan, W.; Parmenter, C.; Fisk, I.D. Stability of Lactobacillus rhamnosus GG incorporated in edible films: Impact of anionic biopolymers and whey protein concentrate. *Food Hydrocoll.* **2017**, *70*, 345–355. [CrossRef] [PubMed]

166. Cecchini, J.P.; Spotti, M.J.; Piagentini, A.M.; Milt, V.G.; Carrara, C.R. Development of edible films obtained from submicron emulsions based on whey protein concentrate, oil/beeswax and brea gum. *Food Sci. Technol. Int.* **2017**, *23*, 371–381. [CrossRef] [PubMed]

167. Ribeiro-Santos, R.; de Melo, N.R.; Andrade, M.; Azevedo, G.; Machado, A.V.; Carvalho-Costa, D.; Sanches-Silva, A. Whey protein active films incorporated with a blend of essential oils: Characterization and effectiveness. *Packag. Technol. Sci.* **2018**, *31*, 27–40. [CrossRef]

168. Pereira, R.C.; Carneiro, J.d.D.S.; Assis, O.B.; Borges, S.V. Mechanical and structural characterization of whey protein concentrate/montmorillonite/lycopene films. *J. Sci. Food Agric.* **2017**, *97*, 4978–4986. [CrossRef]

169. Pérez, L.M.; Piccirilli, G.N.; Delorenzi, N.J.; Verdini, R.A. Effect of different combinations of glycerol and/or trehalose on physical and structural properties of whey protein concentrate-based edible films. *Food Hydrocoll.* **2016**, *56*, 352–359. [CrossRef]

170. Ganiari, S.; Choulitoudi, E.; Oreopoulou, V. Edible and active films and coatings as carriers of natural antioxidants for lipid food. *Trends Food Sci. Technol.* **2017**, *68*, 70–82. [CrossRef]

171. Andrade, M.A.; Ribeiro-Santos, R.; Costa Bonito, M.C.; Saraiva, M.; Sanches-Silva, A. Characterization of rosemary and thyme extracts for incorporation into a whey protein based film. *LWT* **2018**, *92*, 497–508. [CrossRef]

172. Ribeiro-Santos, R.; Andrade, M.; de Melo, N.R.; dos Santos, F.R.; Neves, I.d.A.; de Carvalho, M.G.; Sanches-Silva, A. Biological activities and major components determination in essential oils intended for a biodegradable food packaging. *Ind. Crop. Prod.* **2017**, *97*, 201–210. [CrossRef]

173. Nicolai, T. Formation and functionality of self-assembled whey protein microgels. *Colloids Surf. B Biointerfaces* **2016**, *137*, 32–38. [CrossRef] [PubMed]

174. Abaee, A.; Madadlou, A. Niosome-loaded cold-set whey protein hydrogels. *Food Chem.* **2016**, *196*, 106–113. [CrossRef] [PubMed]

175. Banerjee, S.; Bhattacharya, S. Food Gels: Gelling Process and New Applications. *Crit. Rev. Food Sci. Nutr.* **2012**, *52*, 334–346. [CrossRef] [PubMed]

176. Nguyen, B.T.; Chassenieux, C.; Nicolai, T.; Schmitt, C. Effect of the pH and NaCl on the microstructure and rheology of mixtures of whey protein isolate and casein micelles upon heating. *Food Hydrocoll.* **2017**, *70*, 114–122. [CrossRef]

177. Kharlamova, A.; Chassenieux, C.; Nicolai, T. Acid-induced gelation of whey protein aggregates: Kinetics, gel structure and rheological properties. *Food Hydrocoll.* **2018**, *81*, 263–272. [CrossRef]

178. Lam, C.W.Y.; Ikeda, S. Physical Properties of Heat-induced Whey Protein Aggregates Formed at pH 5.5 and 7.0. *Food Sci. Technol. Res.* **2017**, *23*, 595–601. [CrossRef]

179. Lazidis, A.; Hancocks, R.D.; Spyropoulos, F.; Kreuß, M.; Berrocal, R.; Norton, I.T. Whey protein fluid gels for the stabilisation of foams. *Food Hydrocoll.* **2016**, *53*, 209–217. [CrossRef]

180. Alavi, F.; Momen, S.; Emam-Djomeh, Z.; Salami, M.; Moosavi-Movahedi, A.A. Radical cross-linked whey protein aggregates as building blocks of non-heated cold-set gels. *Food Hydrocoll.* **2018**, *81*, 429–441. [CrossRef]

181. Kharlamova, A.; Nicolai, T.; Chassenieux, C. Calcium-induced gelation of whey protein aggregates: Kinetics, structure and rheological properties. *Food Hydrocoll.* **2018**, *79*, 145–157. [CrossRef]

182. Ren, F.; Dong, D.; Yu, B.; Hou, Z.-h.; Cui, B. Rheology, thermal properties, and microstructure of heat-induced gel of whey protein–acetylated potato starch. *Starch Stärke* **2017**, *69*, 1600344. [CrossRef]

183. Li, Q.; Zhao, Z. Interaction between lactoferrin and whey proteins and its influence on the heat-induced gelation of whey proteins. *Food Chem.* **2018**, *252*, 92–98. [CrossRef] [PubMed]

184. Selig, M.J.; Dar, B.N.; Kierulf, A.; Ravanfar, R.; Rizvi, S.S.H.; Abbaspourrad, A. Modulation of whey protein-kappa carrageenan hydrogel properties via enzymatic protein modification. *Food Funct.* **2018**, *9*, 2313–2319. [CrossRef] [PubMed]

185. Moayyedi, M.; Eskandari, M.H.; Rad, A.H.E.; Ziaee, E.; Khodaparast, M.H.H.; Golmakani, M.-T. Effect of drying methods (electrospraying, freeze drying and spray drying) on survival and viability of microencapsulated *Lactobacillus rhamnosus* ATCC 7469. *J. Funct. Foods* **2018**, *40*, 391–399. [CrossRef]

186. Sogut, E.; Ili Balqis, A.M.; Nur Hanani, Z.A.; Seydim, A.C. The properties of κ-carrageenan and whey protein isolate blended films containing pomegranate seed oil. *Polym. Test.* **2019**, *77*. [CrossRef]

187. Protte, K.; Weiss, J.; Hinrichs, J.; Knaapila, A. Thermally stabilised whey protein-pectin complexes modulate the thermodynamic incompatibility in hydrocolloid matrixes: A feasibility-study on sensory and rheological characteristics in dairy desserts. *LWT* **2019**, *105*, 336–343. [CrossRef]

188. Rajam, R.; Anandharamakrishnan, C. Microencapsulation of Lactobacillus plantarum (MTCC 5422) with fructooligosaccharide as wall material by spray drying. *LWT Food Sci. Technol.* **2015**, *60*, 773–780. [CrossRef]

189. Su, J.; Wang, X.; Li, W.; Chen, L.; Zeng, X.; Huang, Q.; Hu, B. Enhancing the Viability of *Lactobacillus plantarum* as Probiotics through Encapsulation with High Internal Phase Emulsions Stabilized with Whey Protein Isolate Microgels. *J. Agric. Food Chem.* **2018**, *66*, 12335–12343. [CrossRef]

190. Kwiecień, I.; Kwiecień, M. Application of Polysaccharide-Based Hydrogels as Probiotic Delivery Systems. *Gels* **2018**, *4*, 47. [CrossRef]

191. O'Neill, G.J.; Egan, T.; Jacquier, J.C.; O'Sullivan, M.; Dolores O'Riordan, E. Whey microbeads as a matrix for the encapsulation and immobilisation of riboflavin and peptides. *Food Chem.* **2014**, *160*, 46–52. [CrossRef]

192. Abbasi, A.; Emam-Djomeh, Z.; Mousavi, M.A.E.; Davoodi, D. Stability of vitamin D3 encapsulated in nanoparticles of whey protein isolate. *Food Chem.* **2014**, *143*, 379–383. [CrossRef] [PubMed]

193. Rojas-Moreno, S.; Osorio-Revilla, G.; Gallardo-Velázquez, T.; Cárdenas-Bailón, F.; Meza-Márquez, G. Effect of the cross-linking agent and drying method on encapsulation efficiency of orange essential oil by complex coacervation using whey protein isolate with different polysaccharides. *J. Microencapsul.* **2018**, *35*, 165–180. [CrossRef] [PubMed]

194. Mohammadian, M.; Salami, M.; Momen, S.; Alavi, F.; Emam-Djomeh, Z. Fabrication of curcumin-loaded whey protein microgels: Structural properties, antioxidant activity, and in vitro release behavior. *LWT* **2019**, *103*, 94–100. [CrossRef]

195. Raei, M.; Shahidi, F.; Farhoodi, M.; Jafari, S.M.; Rafe, A. Application of whey protein-pectin nano-complex carriers for loading of lactoferrin. *Int. J. Biol. Macromol.* **2017**, *105*, 281–291. [CrossRef] [PubMed]

196. Assadpour, E.; Maghsoudlou, Y.; Jafari, S.-M.; Ghorbani, M.; Aalami, M. Optimization of folic acid nano-emulsification and encapsulation by maltodextrin-whey protein double emulsions. *Int. J. Biol. Macromol.* **2016**, *86*, 197–207. [CrossRef] [PubMed]

197. Fang, Z.; Bao, H.; Ni, Y.; Choijilsuren, N.; Liang, L. Partition and digestive stability of α-tocopherol and resveratrol/naringenin in whey protein isolate emulsions. *Int. Dairy J.* **2019**, *93*, 116–123. [CrossRef]

198. Nourbakhsh, H.; Madadlou, A.; Emam-Djomeh, Z.; Wang, Y.-C.; Gunasekaran, S.; Mousavi, M.E. One-Pot Procedure for Recovery of Gallic Acid from Wastewater and Encapsulation within Protein Particles. *J. Agric. Food Chem.* **2016**, *64*, 1575–1582. [CrossRef]

199. Jain, A.; Sharma, G.; Ghoshal, G.; Kesharwani, P.; Singh, B.; Shivhare, U.S.; Katare, O.P. Lycopene loaded whey protein isolate nanoparticles: An innovative endeavor for enhanced bioavailability of lycopene and anti-cancer activity. *Int. J. Pharm.* **2018**, *546*, 97–105. [CrossRef]

200. O'Neill, G.J.; Egan, T.; Jacquier, J.C.; O'Sullivan, M.; Dolores O'Riordan, E. Kinetics of immobilisation and release of tryptophan, riboflavin and peptides from whey protein microbeads. *Food Chem.* **2015**, *180*, 150–155. [CrossRef]

201. Alavi, F.; Emam-Djomeh, Z.; Yarmand, M.S.; Salami, M.; Momen, S.; Moosavi-Movahedi, A.A. Cold gelation of curcumin loaded whey protein aggregates mixed with k-carrageenan: Impact of gel microstructure on the gastrointestinal fate of curcumin. *Food Hydrocoll.* **2018**, *85*, 267–280. [CrossRef]

202. Mohammadian, M.; Salami, M.; Momen, S.; Alavi, F.; Emam-Djomeh, Z.; Moosavi-Movahedi, A.A. Enhancing the aqueous solubility of curcumin at acidic condition through the complexation with whey protein nanofibrils. *Food Hydrocoll.* **2019**, *87*, 902–914. [CrossRef]

203. Mohammadian, M.; Madadlou, A. Cold-set hydrogels made of whey protein nanofibrils with different divalent cations. *Int. J. Biol. Macromol.* **2016**, *89*, 499–506. [CrossRef] [PubMed]

204. Mohammadian, M.; Salami, M.; Emam-Djomeh, Z.; Momen, S.; Moosavi-Movahedi, A.A. Gelation of oil-in-water emulsions stabilized by heat-denatured and nanofibrillated whey proteins through ion bridging or citric acid-mediated cross-linking. *Int. J. Biol. Macromol.* **2018**, *120*, 2247–2258. [CrossRef]

205. Mantovani, R.A.; Fattori, J.; Michelon, M.; Cunha, R.L. Formation and pH-stability of whey protein fibrils in the presence of lecithin. *Food Hydrocoll.* **2016**, *60*, 288–298. [CrossRef]

206. Hashemi, B.; Madadlou, A.; Salami, M. Functional and in vitro gastric digestibility of the whey protein hydrogel loaded with nanostructured lipid carriers and gelled via citric acid-mediated crosslinking. *Food Chem.* **2017**, *237*, 23–29. [CrossRef] [PubMed]

207. Zhu, J.; Sun, X.; Wang, S.; Xu, Y.; Wang, D. Formation of nanocomplexes comprising whey proteins and fucoxanthin: Characterization, spectroscopic analysis, and molecular docking. *Food Hydrocoll.* **2017**, *63*, 391–403. [CrossRef]

208. Bamba, B.S.B.; Shi, J.; Tranchant, C.C.; Xue, S.J.; Forney, C.F.; Lim, L.-T.; Xu, W.; Xu, G. Coencapsulation of Polyphenols and Anthocyanins from Blueberry Pomace by Double Emulsion Stabilized by Whey Proteins: Effect of Homogenization Parameters. *Molecules* **2018**, *23*, 2525. [CrossRef] [PubMed]

209. Chotiko, A.; Sathivel, S. Releasing characteristics of anthocyanins extract in pectin–whey protein complex microcapsules coated with zein. *J. Food Sci. Technol.* **2017**, *54*, 2059–2066. [CrossRef]

210. Rocha, J.D.C.G.; Viana, K.W.C.; Mendonca, A.C.; Neves, N.D.A.; Carvalho, A.F.D.; Minim, V.P.R.; Barros, F.A.R.D.; Stringheta, P.C. Protein beverages containing anthocyanins of jabuticaba. *Food Sci. Technol.* **2019**, *39*, 112–119. [CrossRef]

211. Shen, X.; Zhao, C.; Lu, J.; Guo, M. Physicochemical Properties of Whey-Protein-Stabilized Astaxanthin Nanodispersion and Its Transport via a Caco-2 Monolayer. *J. Agric. Food Chem.* **2018**, *66*, 1472–1478. [CrossRef]

212. Ramos, O.L.; Pereira, R.N.; Martins, A.; Rodrigues, R.; Fuciños, C.; Teixeira, J.A.; Pastrana, L.; Malcata, F.X.; Vicente, A.A. Design of whey protein nanostructures for incorporation and release of nutraceutical compounds in food. *Crit. Rev. Food Sci. Nutr.* **2017**, *57*, 1377–1393. [CrossRef] [PubMed]

213. Sun, W.-W.; Yu, S.-J.; Zeng, X.-A.; Yang, X.-Q.; Jia, X. Properties of whey protein isolate–dextran conjugate prepared using pulsed electric field. *Food Res. Int.* **2011**, *44*, 1052–1058. [CrossRef]

214. Perusko, M.; Al-Hanish, A.; Cirkovic Velickovic, T.; Stanic-Vucinic, D. Macromolecular crowding conditions enhance glycation and oxidation of whey proteins in ultrasound-induced Maillard reaction. *Food Chem.* **2015**, *177*, 248–257. [CrossRef] [PubMed]

215. Díaz, O.; Candia, D.; Cobos, Á. Effects of ultraviolet radiation on properties of films from whey protein concentrate treated before or after film formation. *Food Hydrocoll.* **2016**, *55*, 189–199. [CrossRef]

216. Kutzli, I.; Gibis, M.; Baier, S.K.; Weiss, J. Formation of Whey Protein Isolate (WPI)–Maltodextrin Conjugates in Fibers Produced by Needleless Electrospinning. *J. Agric. Food Chem.* **2018**, *66*, 10283–10291. [CrossRef]

217. Zhong, J.; Mohan, S.D.; Bell, A.; Terry, A.; Mitchell, G.R.; Davis, F.J. Electrospinning of food-grade nanofibres from whey protein. *Int. J. Biol. Macromol.* **2018**, *113*, 764–773. [CrossRef]

218. Colín-Orozco, J.; Zapata-Torres, M.; Rodríguez-Gattorno, G.; Pedroza-Islas, R. Properties of Poly (ethylene oxide)/whey Protein Isolate Nanofibers Prepared by Electrospinning. *Food Biophys.* **2015**, *10*, 134–144. [CrossRef]

219. Drosou, C.; Krokida, M.; Biliaderis, C.G. Composite pullulan-whey protein nanofibers made by electrospinning: Impact of process parameters on fiber morphology and physical properties. *Food Hydrocoll.* **2018**, *77*, 726–735. [CrossRef]

220. Mendes, A.C.; Stephansen, K.; Chronakis, I.S. Electrospinning of food proteins and polysaccharides. *Food Hydrocoll.* **2017**, *68*, 53–68. [CrossRef]

221. Vieira da Silva, S.; Sogari Picolotto, R.; Wagner, R.; dos Santos Richards, N.S.P.; Smanioto Barin, J. Elemental (Macro- and Microelements) and Amino Acid Profile of Milk Proteins Commercialized in Brazil and Their Nutritional Value. *J. Food Nutr. Res.* **2015**, *3*, 430–436.

222. Devries, M.C.; Phillips, S.M. Supplemental Protein in Support of Muscle Mass and Health: Advantage Whey. *J. Food Sci.* **2015**, *80*, A8–A15. [CrossRef] [PubMed]

223. Trachootham, D.; Lu, W.; Ogasawara, M.A.; Nilsa, R.-D.V.; Huang, P. Redox regulation of cell survival. *Antioxid Redox Signal* **2008**, *10*, 1343–1374. [CrossRef] [PubMed]

224. Pedroso, J.A.B.; Zampieri, T.T.; Donato, J. Reviewing the Effects of l-Leucine Supplementation in the Regulation of Food Intake, Energy Balance, and Glucose Homeostasis. *Nutrients* **2015**, *7*, 3914–3937. [CrossRef] [PubMed]

225. Nie, C.; He, T.; Zhang, W.; Zhang, G.; Ma, X. Branched Chain Amino Acids: Beyond Nutrition Metabolism. *Int. J. Mol. Sci.* **2018**, *19*, 954. [CrossRef] [PubMed]

226. Moura, C.S.; Lollo, P.C.B.; Morato, P.N.; Risso, E.M.; Amaya-Farfan, J. Bioactivity of food peptides: Biological response of rats to bovine milk whey peptides following acute exercise. *Food Nutr. Res.* **2017**, *61*, 1290740. [CrossRef] [PubMed]

227. Ikwegbue, P.C.; Masamba, P.; Oyinloye, B.E.; Kappo, A.P. Roles of Heat Shock Proteins in Apoptosis, Oxidative Stress, Human Inflammatory Diseases, and Cancer. *Pharmaceuticals* **2018**, *11*, 2. [CrossRef] [PubMed]

228. McPherson, R.A.; Hardy, G. Clinical and nutritional benefits of cysteine-enriched protein supplements. *Curr. Opin. Clin. Nutr. Metab. Care* **2011**, *14*, 562–568. [CrossRef]

229. Winter, A.N.; Ross, E.K.; Daliparthi, V.; Sumner, W.A.; Kirchhof, D.M.; Manning, E.; Wilkins, H.M.; Linseman, D.A. A Cystine-Rich Whey Supplement (Immunocal(R)) Provides Neuroprotection from Diverse Oxidative Stress-Inducing Agents In Vitro by Preserving Cellular Glutathione. *Oxid. Med. Cell. Longev.* **2017**, *2017*, 3103272. [CrossRef]

230. Gao, X.; Sanderson, S.M.; Dai, Z.; Reid, M.A.; Cooper, D.E.; Lu, M.; Richie, J.P.; Ciccarella, A.; Calcagnotto, A.; Mikhael, P.G.; et al. Dietary methionine restriction targets one carbon metabolism in humans and produces broad therapeutic responses in cancer. *bioRxiv* **2019**, 627364. [CrossRef]

231. Zheng, G.; Liu, H.; Zhu, Z.; Zheng, J.; Liu, A. Selenium modification of β-lactoglobulin (β-Lg) and its biological activity. *Food Chem.* **2016**, *204*, 246–251. [CrossRef]

232. Sah, B.N.P.; McAinch, A.J.; Vasiljevic, T. Modulation of bovine whey protein digestion in gastrointestinal tract: A comprehensive review. *Int. Dairy J.* **2016**, *62*, 10–18 [CrossRef]

233. Corrêa, A.P.F.; Daroit, D.J.; Fontoura, R.; Meira, S.M.M.; Segalin, J.; Brandelli, A. Hydrolysates of sheep cheese whey as a source of bioactive peptides with antioxidant and angiotensin-converting enzyme inhibitory activities. *Peptides* **2014**, *61*, 48–55. [CrossRef] [PubMed]

234. El-Desouky, W.I.; Mahmoud, A.H.; Abbas, M.M. Antioxidant potential and hypolipidemic effect of whey protein against gamma irradiation induced damages in rats. *Appl. Radiat. Isot.* **2017**, *129*, 103–107. [CrossRef] [PubMed]

235. Yao, C.K.; Muir, J.G.; Gibson, P.R. Review article: Insights into colonic protein fermentation, its modulation and potential health implications. *Aliment. Pharmacol. Ther.* **2016**, *43*, 181–196. [CrossRef] [PubMed]

236. Fonseca, D.P.; Khalil, N.M.; Mainardes, R.M. Bovine serum albumin-based nanoparticles containing resveratrol: Characterization and antioxidant activity. *J. Drug Deliv. Sci. Technol.* **2017**, *39*, 147–155. [CrossRef]

237. Park, Y.W.; Nam, M.S. Bioactive Peptides in Milk and Dairy Products: A Review. *Korean J. Food Sci. Anim. Resour.* **2015**, *35*, 831–840. [CrossRef] [PubMed]

238. Lagrange, V.; Clark, D.C. Chapter 15—Nutritive and Therapeutic Aspects of Whey Proteins. In *Whey Proteins*; Deeth, H.C., Bansal, N., Eds.; Academic Press: Cambridge, MA, USA, 2019; pp. 549–577.

239. Tong, X.; Li, W.; Xu, J.-Y.; Han, S.; Qin, L.-Q. Effects of whey protein and leucine supplementation on insulin resistance in non-obese insulin-resistant model rats. *Nutrition* **2014**, *30*, 1076–1080. [CrossRef]

240. Akhavan, T.; Luhovyy, B.L.; Panahi, S.; Kubant, R.; Brown, P.H.; Anderson, G.H. Mechanism of action of pre-meal consumption of whey protein on glycemic control in young adults. *J. Nutr. Biochem.* **2014**, *25*, 36–43. [CrossRef]

241. Bamdad, F.; Bark, S.; Kwon, C.H.; Suh, J.-W.; Sunwoo, H. Anti-Inflammatory and Antioxidant Properties of Peptides Released from β-Lactoglobulin by High Hydrostatic Pressure-Assisted Enzymatic Hydrolysis. *Molecules* **2017**, *22*, 949. [CrossRef]

242. Alvarado, Y.; Muro, C.; Illescas, J.; Díaz, M.d.C.; Riera, F. Encapsulation of Antihypertensive Peptides from Whey Proteins and Their Releasing in Gastrointestinal Conditions. *Biomolecules* **2019**, *9*, 164. [CrossRef]

243. Tahavorgar, A.; Vafa, M.; Shidfar, F.; Gohari, M.; Heydari, I. Whey protein preloads are more beneficial than soy protein preloads in regulating appetite, calorie intake, anthropometry, and body composition of overweight and obese men. *Nutr. Res.* **2014**, *34*, 856–861. [CrossRef] [PubMed]

244. Tahavorgar, A.; Vafa, M.; Shidfar, F.; Gohari, M.; Heydari, I. Beneficial effects of whey protein preloads on some cardiovascular diseases risk factors of overweight and obese men are stronger than soy protein preloads—A randomized clinical trial. *J. Nutr. Intermed. Metab.* **2015**, *2*, 69–75. [CrossRef]

245. Lollo, P.C.B.; Amaya-Farfan, J.; Faria, I.C.; Salgado, J.V.V.; Chacon-Mikahil, M.P.T.; Cruz, A.G.; Oliveira, C.A.F.; Montagner, P.C.; Arruda, M. Hydrolysed whey protein reduces muscle damage markers in Brazilian elite soccer players compared with whey protein and maltodextrin. A twelve-week in-championship intervention. *Int. Dairy J.* **2014**, *34*, 19–24. [CrossRef]

246. Cheung, L.K.Y.; Aluko, R.E.; Cliff, M.A.; Li-Chan, E.C.Y. Effects of exopeptidase treatment on antihypertensive activity and taste attributes of enzymatic whey protein hydrolysates. *J. Funct. Foods* **2015**, *13*, 262–275. [CrossRef]

247. Kimura, Y.; Sumiyoshi, M.; Kobayashi, T. Whey Peptides Prevent Chronic Ultraviolet B Radiation–Induced Skin Aging in Melanin-Possessing Male Hairless Mice. *J. Nutr.* **2013**, *144*, 27–32. [CrossRef]

248. Dalziel, J.E.; Anderson, R.C.; Bassett, S.A.; Lloyd-West, C.M.; Haggarty, N.W.; Roy, N.C. Influence of Bovine Whey Protein Concentrate and Hydrolysate Preparation Methods on Motility in the Isolated Rat Distal Colon. *Nutrients* **2016**, *8*, 809. [CrossRef]

249. Nilaweera, K.N.; Cabrera-Rubio, R.; Speakman, J.R.; O'Connor, P.M.; McAuliffe, A.; Guinane, C.M.; Lawton, E.M.; Crispie, F.; Aguilera, M.; Stanley, M.; et al. Whey protein effects on energy balance link the intestinal mechanisms of energy absorption with adiposity and hypothalamic neuropeptide gene expression. *Am. J. Physiol. Endocrinol. Metab.* **2017**, *313*, E1–E11. [CrossRef]

250. Hwang, J.S.; Han, S.G.; Lee, C.H.; Seo, H.G. Whey Protein Attenuates Angiotensin II-Primed Premature Senescence of Vascular Smooth Muscle Cells through Upregulation of SIRT1. *Korean J. Food Sci. Anim. Resour.* **2017**, *37*, 917–925. [CrossRef]

251. Garg, G.; Singh, S.; Singh, A.K.; Rizvi, S.I. Whey protein concentrate supplementation protects rat brain against aging-induced oxidative stress and neurodegeneration. *Appl. Physiol. Nutr. Metab.* **2017**, *43*, 437–444. [CrossRef]

252. Flaim, C.; Kob, M.; Di Pierro, A.M.; Herrmann, M.; Lucchin, L. Effects of a whey protein supplementation on oxidative stress, body composition and glucose metabolism among overweight people affected by diabetes mellitus or impaired fasting glucose: A pilot study. *J. Nutr. Biochem.* **2017**, *50*, 95–102. [CrossRef]

253. Ney, D.M.; Blank, R.D.; Hansen, K.E. Advances in the nutritional and pharmacological management of phenylketonuria. *Curr. Opin. Clin. Nutr. Metab. Care* **2014**, *17*, 61–68. [CrossRef] [PubMed]

254. Brown, M.A.; Stevenson, E.J.; Howatson, G. Whey protein hydrolysate supplementation accelerates recovery from exercise-induced muscle damage in females. *Appl. Physiol. Nutr. Metab.* **2017**, *43*, 324–330. [CrossRef] [PubMed]

255. Shirato, M.; Tsuchiya, Y.; Sato, T.; Hamano, S.; Gushiken, T.; Kimura, N.; Ochi, E. Effects of combined β-hydroxy-β-methylbutyrate (HMB) and whey protein ingestion on symptoms of eccentric exercise-induced muscle damage. *J. Int. Soc. Sports Nutr.* **2016**, *13*, 7. [CrossRef] [PubMed]

256. Moura, C.S.; Lollo, P.C.B.; Morato, P.N.; Nisishima, L.H.; Carneiro, E.M.; Amaya-Farfan, J. Whey Protein Hydrolysate Enhances HSP90 but Does Not Alter HSP60 and HSP25 in Skeletal Muscle of Rats. *PLoS ONE* **2014**, *9*, e83437. [CrossRef]

257. Korhonen, H.; Pihlanto, A. Bioactive peptides: Production and functionality. *Int. Dairy J.* **2006**, *16*, 945–960. [CrossRef]

258. Muro Urista, C.; Alvarez Fernandez, R.; Riera Rodriguez, F.; Arana Cuenca, A.; Tellez Jurado, A. Review: Production and functionality of active peptides from milk. *Food Sci. Technol. Int.* **2011**, *17*, 293–317. [CrossRef]

259. Welsh, G.; Ryder, K.; Brewster, J.; Walker, C.; Mros, S.; Bekhit, A.E.-D.A.; McConnell, M.; Carne, A. Comparison of bioactive peptides prepared from sheep cheese whey using a food-grade bacterial and a fungal protease preparation. *Int. J. Food Sci. Technol.* **2017**, *52*, 1252–1259. [CrossRef]

260. Rocha, G.F.; Kise, F.; Rosso, A.M.; Parisi, M.G. Potential antioxidant peptides produced from whey hydrolysis with an immobilized aspartic protease from *Salpichroa origanifolia* fruits. *Food Chem.* **2017**, *237*, 350–355. [CrossRef]

261. Hernández-Ledesma, B.; Hsieh, C.-C.; Martínez-Villaluenga, C. Food Bioactive Compounds against Diseases of the 21st Century 2016. *BioMed Res. Int.* **2017**, *2017*. [CrossRef]

262. Brandelli, A.; Daroit, D.J.; Corrêa, A.P.F. Whey as a source of peptides with remarkable biological activities. *Food Res. Int.* **2015**, *73*, 149–161. [CrossRef]

263. Brumini, D.; Criscione, A.; Bordonaro, S.; Vegarud, G.E.; Marletta, D. Whey proteins and their antimicrobial properties in donkey milk: A brief review. *Dairy Sci. Technol.* **2016**, *96*, 1–14. [CrossRef]

264. Mohan, A.; Udechukwu, M.C.; Rajendran, S.R.C.K.; Udenigwe, C.C. Modification of peptide functionality during enzymatic hydrolysis of whey proteins. *RSC Adv.* **2015**, *5*, 97400–97407. [CrossRef]

265. Adams, R.L.; Broughton, K.S. Insulinotropic Effects of Whey: Mechanisms of Action, Recent Clinical Trials, and Clinical Applications. *Ann. Nutr. Metab.* **2016**, *69*, 56–63. [CrossRef] [PubMed]

266. Dullius, A.; Goettert, M.I.; de Souza, C.F.V. Whey protein hydrolysates as a source of bioactive peptides for functional foods—Biotechnological facilitation of industrial scale-up. *J. Funct. Foods* **2018**, *42*, 58–74. [CrossRef]

267. Agyei, D.; Ongkudon, C.M.; Wei, C.Y.; Chan, A.S.; Danquah, M.K. Bioprocess challenges to the isolation and purification of bioactive peptides. *Food Bioprod. Process.* **2016**, *98*, 244–256. [CrossRef]

268. Iltchenco, S.; Preci, D.; Bonifacino, C.; Fraguas, E.F.; Steffens, C.; Panizzolo, L.A.; Colet, R.; Fernandes, I.A.; Abirached, C.; Valduga, E.; et al. Whey protein concentration by ultrafiltration and study of functional properties. *Ciênc. Rural* **2018**, *48*. [CrossRef]

269. Nongonierma, A.B.; FitzGerald, R.J. Strategies for the discovery and identification of food protein-derived biologically active peptides. *Trends Food Sci. Technol.* **2017**, *69*, 289–305. [CrossRef]

270. Dupont, D. Peptidomic as a tool for assessing protein digestion. *Curr. Opin. Food Sci.* **2017**, *16*, 53–58. [CrossRef]

271. Valdés, A.; Cifuentes, A.; León, C. Foodomics evaluation of bioactive compounds in foods. *TrAC Trends Anal. Chem.* **2017**, *96*, 2–13. [CrossRef]

272. Agyei, D.; Tsopmo, A.; Udenigwe, C.C. Bioinformatics and peptidomics approaches to the discovery and analysis of food-derived bioactive peptides. *Anal. Bioanal. Chem.* **2018**, *410*, 3463–3472. [CrossRef]

273. de Jong, E.; Higson, A.; Walsh, P.; Wellissch, M. Bio-based chemicals: Value added products from biorefineries. 2012. Available online: http://www.iea-bioenergy.task42-biorefineries.com/publications/reports (accessed on 15 June 2019).

274. De Corato, U.; De Bari, I.; Viola, E.; Pugliese, M. Assessing the main opportunities of integrated biorefining from agro-bioenergy co/by-products and agroindustrial residues into high-value added products associated to some emerging markets: A review. *Renew. Sustain. Energy Rev.* **2018**, *88*, 326–346. [CrossRef]

275. Matsakas, L.; Gao, Q.; Jansson, S.; Rova, U.; Christakopoulos, P. Green conversion of municipal solid wastes into fuels and chemicals. *Electron. J. Biotechnol.* **2017**, *26*, 69–83. [CrossRef]

276. Iriondo-DeHond, M.; Miguel, E.; Del Castillo, M.D. Food Byproducts as Sustainable Ingredients for Innovative and Healthy Dairy Foods. *Nutrients* **2018**, *10*, 1358. [CrossRef] [PubMed]

277. Fermoso, F.G.; Serrano, A.; Alonso-Fariñas, B.; Fernández-Bolaños, J.; Borja, R.; Rodríguez-Gutiérrez, G. Valuable Compound Extraction, Anaerobic Digestion, and Composting: A Leading Biorefinery Approach for Agricultural Wastes. *J. Agric. Food Chem.* **2018**, *66*, 8451–8468. [CrossRef] [PubMed]

278. Popa, V.I. 1—Biomass for Fuels and Biomaterials. In *Biomass as Renewable Raw Material to Obtain Bioproducts of High-Tech Value*; Popa, V., Volf, I., Eds.; Elsevier: Amsterdam, The Netherlands, 2018; pp. 1–37.

279. Fu, W.; Mathews, A.P. Lactic acid production from lactose by *Lactobacillus plantarum*: Kinetic model and effects of pH, substrate, and oxygen. *BioChem. Eng. J.* **1999**, *3*, 163–170. [CrossRef]

280. Petrov, K.K.; Yankov, D.S.; Beschkov, V.N. Lactic acid fermentation by cells of *Lactobacillus rhamnosus* immobilized in polyacrylamide gel. *World J. Microbiol. Biotechnol.* **2006**, *22*, 337–345. [CrossRef]

281. Sørensen, K.I.; Curic-Bawden, M.; Junge, M.P.; Janzen, T.; Johansen, E. Enhancing the Sweetness of Yoghurt through Metabolic Remodeling of Carbohydrate Metabolism in *Streptococcus thermophilus* and *Lactobacillus delbrueckii* subsp. *bulgaricus*. *Appl. Environ. Microbiol.* **2016**, *82*, 3683–3692. [CrossRef]

282. Panesar, P.S.; Kennedy, J.F.; Knill, C.J.; Kosseva, M. Production of L(+) lactic acid using *Lactobacillus casei* from whey. *Braz. Arch. Biol. Technol.* **2010**, *53*, 219–226. [CrossRef]

283. Rama, G.R.; Kuhn, D.; Beux, S.; Maciel, M.J.; Volken de Souza, C.F. Potential applications of dairy whey for the production of lactic acid bacteria cultures. *Int. Dairy J.* **2019**, *98*, 25–37. [CrossRef]

284. Amaro, T.M.M.M.; Rosa, D.; Comi, G.; Iacumin, L. Prospects for the Use of Whey for Polyhydroxyalkanoate (PHA) Production. *Front. Microbiol.* **2019**, *10*, 992. [CrossRef]

285. Brown, K.; Harrison, J.; Bowers, K. Production of Oxalic Acid from Aspergillus niger and Whey Permeate. *Water Air Soil Pollut.* **2017**, *229*, 5. [CrossRef]

286. Pasotti, L.; Zucca, S.; Casanova, M.; Micoli, G.; Cusella De Angelis, M.G.; Magni, P. Fermentation of lactose to ethanol in cheese whey permeate and concentrated permeate by engineered *Escherichia coli*. *BMC Biotechnol.* **2017**, *17*, 48. [CrossRef] [PubMed]

287. Zhou, X.; Hua, X.; Huang, L.; Xu, Y. Bio-utilization of cheese manufacturing wastes (cheese whey powder) for bioethanol and specific product (galactonic acid) production via a two-step bioprocess. *Bioresour. Technol.* **2019**, *272*, 70–76. [CrossRef] [PubMed]

288. Venkata Mohan, S.; Nikhil, G.N.; Chiranjeevi, P.; Nagendranatha Reddy, C.; Rohit, M.V.; Kumar, A.N.; Sarkar, O. Waste biorefinery models towards sustainable circular bioeconomy: Critical review and future perspectives. *Bioresour. Technol.* **2016**, *215*, 2–12. [CrossRef] [PubMed]

289. O'Callaghan, K. Technologies for the utilisation of biogenic waste in the bioeconomy. *Food Chem.* **2016**, *198*, 2–11. [CrossRef] [PubMed]

290. Bekatorou, A.; Plioni, I.; Sparou, K.; Maroutsiou, R.; Tsafrakidou, P.; Petsi, T.; Kordouli, E. Bacterial Cellulose Production Using the Corinthian Currant Finishing Side-Stream and Cheese Whey: Process Optimization and Textural Characterization. *Foods* **2019**, *8*, 193. [CrossRef] [PubMed]

291. Kopsahelis, N.; Dimou, C.; Papadaki, A.; Xenopoulos, E.; Kyraleou, M.; Kallithraka, S.; Kotseridis, Y.; Papanikolaou, S.; Koutinas, A.A. Refining of wine lees and cheese whey for the production of microbial oil, polyphenol-rich extracts and value-added co-products. *J. Chem. Technol. Biotechnol.* **2018**, *93*, 257–268. [CrossRef]

292. Dimitrellou, D.; Kandylis, P.; Kourkoutas, Y.; Kanellaki, M. Novel probiotic whey cheese with immobilized lactobacilli on casein. *LWT* **2017**, *86*, 627–634. [CrossRef]

293. Paximada, P.; Koutinas, A.A.; Scholten, E.; Mandala, I.G. Effect of bacterial cellulose addition on physical properties of WPI emulsions. Comparison with common thickeners. *Food Hydrocoll.* **2016**, *54*, 245–254. [CrossRef]

294. Gama, A.P.; Hung, Y.-C.; Adhikari, K. Optimization of Emulsifier and Stabilizer Concentrations in a Model Peanut-Based Beverage System: A Mixture Design Approach. *Foods* **2019**, *8*, 116. [CrossRef]

295. Peng, J.; Calabrese, V.; Geurtz, J.; Velikov, K.P.; Venema, P.; van der Linden, E. Composite Gels Containing Whey Protein Fibrils and Bacterial Cellulose Microfibrils. *J. Food Sci.* **2019**, *84*, 1094–1103. [CrossRef] [PubMed]

296. Bavyko, O.; Bondarchuk, M. Ice-cream with functional properties as a means of commercial networks assortment extension and population feeding improving. *J. Hyg. Eng. Des.* **2019**, *26*, 127–133.

297. Agustini, T.W.; Ma'ruf, W.F.; Widayat, W.; Suzery, M.; Hadiyanto, H.; Benjakul, S. Application of *Spirulina platensis* on ice cream and soft cheese with respect to their nutritional and sensory perspectives. *J. Teknol.* **2016**, *78*, 245–251. [CrossRef]

298. Moreira, J.B.; Lim, L.-T.; da Zavareze, R.E.; Dias, A.R.G.; Costa, J.A.V.; de Morais, M.G. Antioxidant ultrafine fibers developed with microalga compounds using a free surface electrospinning. *Food Hydrocoll.* **2019**, *93*, 131–136. [CrossRef]

299. Batista, A.P.; Nunes, M.C.; Fradinho, P.; Gouveia, L.; Sousa, I.; Raymundo, A.; Franco, J.M. Novel foods with microalgal ingredients—Effect of gel setting conditions on the linear viscoelasticity of *Spirulina* and *Haematococcus* gels. *J. Food Eng.* **2012**, *110*, 182–189. [CrossRef]

300. Gouveia, L. Spirulina maxima and Diacronema vlkianum microalgae in vegetable gelled desserts. *Nutr. Food Sci.* **2008**, *38*, 492–501. [CrossRef]

301. Terpou, A.; Papadaki, A.; Lappa, I.K.; Kachrimanidou, V.; Bosnea, L.A.; Kopsahelis, N. Probiotics in Food Systems: Significance and Emerging Strategies Towards Improved Viability and Delivery of Enhanced Beneficial Value. *Nutrients* **2019**, *11*, 1591. [CrossRef]

MDPI

St. Alban-Anlage 66

4052 Basel

Switzerland

Tel. +41 61 683 77 34

Fax +41 61 302 89 18

www.mdpi.com

Foods Editorial Office

E-mail: foods@mdpi.com

www.mdpi.com/journal/foods

www.ingramcontent.com/pod-product-compliance
Lightning Source LLC
Chambersburg PA
CBHW051908210326
41597CB00033B/6069